Eduardo Mizraji
The Matrix Algebra of Logic

Also of interest

Eduardo Mizraji

The Matrix Algebra of Logic

From Logic Operators to Cellular Automata

DE GRUYTER

Author
Dr Eduardo Mizraji
Sección Biofísica y Biología de Sistemas
Facultad de Ciencias
Universidad de la República
Iguá 4225
11400 Montevideo
Uruguay
mizraj@fcien.edu.uy

ISBN 978-3-11-223004-6
e-ISBN (PDF) 978-3-11-223005-3
e-ISBN (EPUB) 978-3-11-223006-0

Library of Congress Cataloging-in-Publication Data
A CIP catalog record for this book has been applied for at the Library of Congress.

Bibliographic information published by the Deutsche Nationalbibliothek
The Deutsche Nationalbibliothek lists this publication in the Deutsche Nationalbibliografie;
detailed bibliographic data are available on the Internet at http://dnb.dnb.de.

© 2026 Walter de Gruyter GmbH, Berlin/Boston, Genthiner Straße 13, 10785 Berlin
Cover image: Just_Super/E+/Getty Images
Typesetting: Integra Software Services Pvt. Ltd.

www.degruyterbrill.com
Questions about General Product Safety Regulation:
productsafety@degruyterbrill.com

To Cristina, who is still by my side

Acknowledgments

In the European summer of 1976, Cristina Arruti was temporarily staying with relatives in Galicia, prior to our scheduled meeting in September in Paris, where we both had our respective scholarships. During that Galician stay, Cristina, captivated by logic after Mario Otero's long-standing lectures in Montevideo, bought Jan Lukasiewicz's *Estudios de Lógica y Filosofía* (*Studies in Logic and Philosophy*) in Pontevedra, published by *Revista de Occidente*. It was not foreseeable at the time that this book would inspire and guide me many years later in my explorations in the field of logic. These explorations led to the results presented in this book. To my dear Cristina, then, my first thanks go, for this and for so many things that have illuminated my life.

I am grateful to my colleagues and friends Andrés Pomi, Álvaro Cabana, Juan Lin, and Juan Valle Lisboa for their continued interest and encouragement in my work on the vector formalism of logic.

I also thank the Faculty of Sciences of the University of the Republic for the broad freedom and comfortable research environment it has provided me since its founding. This book was written during a sabbatical year granted by the Faculty of Sciences.

I thank Álvaro Risso and his team for their editorial assistance with the original Spanish edition. Finally, I would like to thank Damiano Sacco and Melanie Götz of De Gruyter GmbH, Berlin for their continued support in the publication of this book.

https://doi.org/10.1515/9783112230053-202

Preface

The search for the origin of human rationality has been an epic journey that, consolidated by the immortal contributions of ancient Greece, has continued through the centuries, adopting the ideas and technical instruments that science has contributed. In the nineteenth century, inspired by the dazzling achievements of the physical sciences, great innovators such as Babbage, De Morgan, and especially Boole took on the adventure of trying to represent the laws of human reasoning with mathematical techniques inspired by physics and astronomy. Already in the twentieth century, attempts were made to use the same type of epistemology to understand how the intricate networks of neurons in the human brain were able to support our cognitive life, including memories, concept creation, language generation, and the dynamics of thought. This led to the creation of various mathematical models to capture the complexity of neural networks. It is within this framework that the discovery to which this book is dedicated arose.

While researching a model of neural memory in 1987, I decided to explore whether that model could represent a logical operation called exclusive-or (XOR). The model used represented a slight innovation compared to associative memory models based on linear algebra formalisms, which had been extensively researched since the early 1970s. Since the pioneering model for neural networks created by McCulloch and Pitts in 1943, the logical function XOR had shown difficulties in representation, which contrasted with the ease of computation in that model of other operations, such as disjunction or implication. Similarly, other later neural models, including the matrix memory models of the 1970s, had structural incapacities to compute the XOR in a simple way.

For reasons that will be briefly discussed in Chapter 3, the model I was investigating included vector contexts that were associated with memory entries through a tensor operation, the Kronecker product. Testing the ability of this neural memory model to compute XOR yielded a surprising result: the representation of the operation was immediate, and the XOR computation was performed in a single step.

Let us note that the second half of the 1980s was a time when computing the XOR was considered an important test for assessing the contextualization capacity of neurocomputational models. This search for neuromimetic ways of computing the XOR had produced one of the most influential neural models in the history of the subject: the neural network model with hidden layers and nonlinear connections. And alongside this model emerged one of the most powerful and widely used algorithms of our time: the method of training neural networks through backpropagation of errors with the aim of minimizing them.

The easy computation of XOR in my model contrasted with the climate of algorithmic complexity that was permeating the field of research into neural memory models, with the added advantage that my neural model had relatively simple mathematics,

https://doi.org/10.1515/9783112230053-203

which allowed for further theoretical studies (which were not easy with the formalism of nonlinear hidden layer memories).

Having completed the initial stage of this investigation into a context-modulated memory model (which fortunately was able to be transformed into a paper published in 1989), I became curious about the meaning and scope of this simple representation of XOR. This curiosity led me to discover that there was a formalism based on linear algebra that allowed the representation of a large region of logic. I knew that in the field of quantum physics a matrix formalism for logic had emerged known as "quantum logic," but the aspirations of my approach were more modest, as it merely involved exploring a "translation" of elementary logic into a new format. But surprises arose along the way, and some will be recounted in this book.

The formalism presented here has another inspiration. Born from biology, it is intended as a modest continuation of the program of symbolic representation of logical operations as the basis of rational thought, initiated by George Boole when he published *The Mathematical Analysis of Logic* in 1847 and continued from another perspective by the research of the Polish logician Jan Lukasiewicz.

Eduardo Mizraji

Contents

Introduction

The term "logicism" refers to a doctrine according to which logic underpins mathematics and from it derives the rules that make proofs valid. Within this framework, logic is the foundation that supports the vast edifice of mathematics. The origin of this doctrine is generally attributed to Peano and Frege, but its consolidation was almost certainly due to the intellectual vigor and influence of Bertrand Russell, and especially to the monumental work *Principia Mathematica*, which he published with Alfred North Whitehead between 1910 and 1913 (Rodríguez-Consuegra 1994).

George Boole is a name most associated today with the binary logic used in digital computing. This current binary logic is based on algebraic operations that are a modified version of those originally created by Boole (1847, 1854 [reprinted in 1958]). The algebraic approach to logical operations was developed during the nineteenth century and is referred to as "Algebraic Logic" in the taxonomy of the history of mathematics (Houser 1994).

The protagonists in the creation of algebraic logic are primarily George Boole, Augustus De Morgan, Stanley Jevons, Charles Peirce, Ernst Schöreder, and Alfred N. Whitehead, and their contributions span from 1840 to 1900.

Considering the intense influence of logicism on mathematics, it gives the impression that algebraic logic was a kind of slow, clumsy caterpillar that then marvelously evolved at the dawn of the twentieth century into a beautiful, long-range butterfly that became contemporary mathematical logic. That this is a valid metaphor is attested by its extraordinary impact on all branches of science and technology.

There is a logicist reinterpretation of Boole's work that assumes that it had nothing to do with thought, or that thought was really outside of Boole's interests. Fortunately, Boole's two books on logic are immediately accessible to anyone who wants to consult them (Boole 1847, 1854), and they will see that this reinterpretation contradicts the aims explicitly stated by Boole, especially in his *The Laws of Thought* of 1854. In addition, the subtitle of *The Mathematical Analysis of Logic*, his 1847 book, is *Being an Essay Towards a Calculus of Deductive Reasoning*.

It is clear, however, that Boole did not intend to cover human psychology but to focus on that area of thought on which the emergence of rationality depended.

Getting closer to the objective of this book, let us note the following. Leibniz's representations for differential and integral calculus introduced an operator formalism in which the symbols for differentiation d/dx and integration $\int dx$ were objects that could themselves be subject to algebraic operations such as potentiation or scalar multiplication. This notion was powerfully used and developed by mathematicians such as Lagrange, Laplace, or Arbogast, and by the nineteenth century it was adopted by a limited number of British researchers, who emphasized the depth of the operator approach as opposed to Newton's notation, which prevailed in the British Isles. As they said at the time with humor, it was the opposition between the "dot-ist" Newto-

https://doi.org/10.1515/9783112230053-001

nians against the "de-ist" Leibnizians. Among these "de-ist" British mathematicians were the prestigious Charles Babbage and the young Duncan Gregory. Despite lacking formal studies, George Boole, a well-informed autodidact, was fascinated by the works of Lagrange and soon enrolled in the investigation of the mathematical potential of differential calculus operators. This led him to become an expert in the use of symbolic methods in the field of differential equations. At the same time, Boole was intensely interested in the nature of rational thought, and this led him to create the algebraic ideas and techniques that he included in his 1847 book.

Our purpose here is to represent the algebra of logic through a theory of operators. The operators used are matrices that operate on vector truth values. These matrices allow us to represent virtually all operations of propositional calculus and prove its fundamental theorems, but operating according to the rules of matrix algebra. Likewise, various aspects of non-binary logic and modal logic can also be analyzed using this method of algebraic operators. It is important to note that in mathematical logic, an algebraic structure called a "matrix" is defined. However, these formal matrices are not the matrix operators of linear algebra that we will refer to in this book. For details on the formal theory of logical matrices, see chapter 3 of Malinowski's book (1993).

It is important to mention the pioneering contributions of the great biophysicist G.N. Ramachandran, who used matrix algebra to represent the formal structure of various aspects of traditional Indian logic. A complete description of Ramachandra's work on logic and its references can be found in Jain (2011).

Let us note two interesting points before concluding this Introduction. Matrices were formally created by Arthur Cayley in his classic article of 1858. We don't know whether Boole, who died in 1864, ever knew Cayley's matrix formalism (Cayley 1858). There was a tool there to make the algebra of Boolean logic also an algebra of operators. Boole and Cayley knew each other. Boole sent Cayley copy of his 1847 book and a brief correspondence ensued between the two (this correspondence is published in Grattan-Guinness and Bornet 1997).

But the one who did use an operator method for logical calculus was Jan Lukasiewicz when he invented his Polish notation. Paradoxically, Lukasiewicz was part of the tradition of logicism and did not appear to be an algebraist of logic. Yet, at the same time, he was part of an extremely original "insular" group that created what is now called "Polish logic" (Simons 1994). The goal of Polish notation was to be able to write logical expressions without using parentheses, and Lukasiewicz showed that this was always possible. For example, in Polish notation, the operation $a \times (b + c)$ is written $\times a + bc$, and a logical expression like $p \Rightarrow (q \wedge r)$ is written $\Rightarrow p \wedge qr$. The original Polish notation uses literal symbols, which are more convenient than these classic logical symbols. The reader will later see that the matrix formalism presented here, although it does not dispense with parentheses, due to the rules of matrix algebra exactly reproduces the order of operators and variables in Polish notation.

References

Boole, G. (1847) The Mathematical Analysis of Logic, Macmillan, Barclay, and Macmillan, Cambridge, London.

Boole, G. (1854, 1958) An Investigation of the Laws of Thought, Dover, New York.

Cayley, A. (1858) A Memoir on the Theory of Matrices, Philosophical Transactions of the Royal Society, 148: 17–37.

Grattan-Guinness, I.Y., Bornet, G. (1997) George Boole, in Selected Manuscripts in Logic and Philosophy, Birkhäuser, Basel.

Houser, N. (1994) Algebraic logic from Boole to Schröder, in "Companion Encyclopedia of the History and Phylosophy of the Mathematical Sciences, (I. Grattan-Guinness, Ed), Routledge, London.

Jain, M.K. (2011) Logic of evidence-based inference propositions. Current Science, 100, 1663–1672.

Malinowski, G. (1993) Many-Valued Logics, Clarendon Press, Oxford.

Rodríguez-Consuegra, F.A. (1994) Mathematical logic and Logicism from Peano to Quine, in "Companion Encyclopedia of the History and Phylosophy of the Mathematical Sciences, (I. Grattan-Guinness, Ed), Routledge, London.

Simons, P. (1994) Polish Logic, in "Companion Encyclopedia of the History and Phylosophy of the Mathematical Sciences, (I. Grattan-Guinness, Ed), Routledge, London.

Part 1: **The foundations**

Chapter 1
The bases of logical calculation

We will begin by assuming that a central component of intellectual interactions between people is based on the reception and understanding of propositions that must be explicitly or implicitly judged. For example, one scorching day someone in an elevator receives the comment "this heat is unbearable," which for him is true. On another occasion, a friend tells him that the girl they are coming to see is quite ugly, but he judges this to be false. And yet another day, he is told that one hundred times one hundred equals one thousand, which for him is a falsehood.

Let us then first define an abstract set of propositions that an individual could receive:

$$\wp = \{p_1, p_2, \ldots, p_\omega\}. \tag{1.1}$$

The p_i $(i = 1, \ldots, \omega)$ are objects of the language used (verbal as in speech, literal as in a text, gestural as in sign language, symbolic as in a mathematical expression, etc.) that transfer information and have the structure of a proposition. This indicates that they can be evaluated as to whether they are true or false. We now define a set τ that contains the appropriately encoded "true" and "false" assignments attributable to a proposition:

$$\tau = \{t, f\}. \tag{1.2}$$

This set of truth values consists of two elements: t = "true" and f = "false." This set will be expanded in later developments. Assuming then that a proposition is a statement to which a truth value can be assigned, the $\mathfrak{J}$ "judgment" application arises:

$$\mathfrak{J} : \wp \to \tau. \tag{1.3}$$

What kinds of objects are the values t and f? Different representations of formal logic define them in different ways. The symbols t and f can be the words "true" and "false," respectively; they can be the numbers 1 and 0; they can represent sets where t is similar to a universal set Y (containing all truths) and f is similar to the empty set ϕ (containing no truths); and we will see later that they can also be vectors.

Usually, the assignment of sets as truth values corresponds to the so-called "class logic," and when the truth variables are not sets, we speak of "propositional calculus." In a famous text by Lukasiewicz (1934) on the history of propositional calculus, the interested reader can find information regarding the differences between the two representations. However, in this exploratory essay, we deliberately aim to maintain an ambiguous link between propositional logic and class logic, since the operator formalism allows for a practical fusion of both approaches. We should note in passing that the algebraic formalism developed by Boole corresponds to a class logic.

https://doi.org/10.1515/9783112230053-002

1.1 Basic logical operations

We will say here that logical functions arise when various connections between propositions are assigned truth values that judge the value of those connections. If it is stated that living humans breathe (which is true) and that granite rocks float in the air (which is false), the sentence "living humans breathe and granite rocks float in the air" is a connection whose truth value is false. That is, the composition of the truth values of two propositions using the word "and" produces another truth value (here: "true and false = false"). Let us represent this symbolically as follows:

$$\Psi\left(p_{j(1)}, \ldots, p_{j(r)}\right) \rightarrow \Phi\left(v_{j(1)}, \ldots, v_{j(r)}\right) \tag{1.4}$$

with $p_{j(i)} \in \wp$ and $v_{j(i)} \in \tau$. The final determination of the truth value of the set of propositions Ψ arises from the links between the truth values that make up the function Φ. It is well known that the function $\Phi\left(v_{j(1)}, \ldots, v_{j(r)}\right)$ can be decomposed into elementary functions of at most two variables. For this reason, it is sufficient to focus on the study of monadic and dyadic (also called "connective") functions. We define a monadic function as follows:

$$\mu : \tau \rightarrow \tau. \tag{1.5}$$

A dyadic function results from the application

$$\delta : \tau \times \tau \rightarrow \tau, \tag{1.6}$$

where $\times$ represents the Cartesian product. A logical operator is thus a monadic or dyadic function defined according to applications (1.5) and (1.6). Note that since there are two elements of τ, there are 4 monadic and 16 dyadic functions. Of the 16 dyadic only one restricted set is used.

1.1.1 Tabular representations

The representation of monadic and dyadic functions by means of the so-called "truth tables" is in common use. To describe them, we will adopt the numerical representation of truth values: $t \mapsto 1$ and $f \mapsto 0$. On the contrary, the propositions to which these values are assigned will be represented by lowercase p, q, r, etc. The origin of this tabular representation is still a matter of debate among historians, and although several logicians used this type of representation (Peirce, for example), there are those who attribute its current format to Ludwig Wittgenstein, the famous student of Bertrand Russell, who expounded them in section 5.101 of the (Wittgenstein 2021). In Figure 1.1, we reproduce the image of section 5.101.

TRACTATUS LOGICO-PHILOSOPHICUS

5.101   The truth-functions of every number of elementary propositions
        can be written in a schema of the following kind:

(TTTT)(p,q) Tautology   (if p then p, and if q then q) $[p \supset p . q \supset q]$
(FTTT)(p,q) in words: Not both p and q. $[\sim(p . q)]$
(TFTT)(p,q) „   „   If q then p. $[q \supset p]$
(TTFT)(p,q) „   „   If p then q. $[p \supset q]$
(TTTF)(p,q) „   „ p or q. $[p \lor q]$
(FFTT)(p,q) „   „   Not q. $[\sim q]$
(FTFT)(p,q) „   „   Not p. $[\sim p]$
(FTTF)(p,q) „   „ p or q, but not both. $[p . \sim q : \lor : q . \sim p]$
(TFFT)(p,q) „   „   If p, then q; and if q, then p. $[p \equiv q]$
(TFTF)(p,q) „   „ p
(TTFF)(p,q) „   „ q
(FFFT)(p,q) „   „   Neither p nor q. $[\sim p . \sim q$ or $p \mid q]$
(FFTF)(p,q) „   „ p and not q. $[p . \sim q]$
(FTFF)(p,q) „   „ q and not p. $[q . \sim p]$
(TFFF)(p,q) „   „ p and q. $[p . q]$
(FFFF)(p,q) Contradiction (p and not p; and q and not q.) $[p . \sim p . q . \sim q]$

Those truth-possibilities of its truth-arguments, which verify
the proposition, I shall call its *truth-grounds*.

Figure 1.1: Dyadic truth tables in Wittgenstein's *Tractaus Logico-Philosophicus* [from Gutenberg Project online edition]. It is suggestive, as shown later (e.g., in eq. (4.15)), that this Wittgenstein tabular format corresponds to one of the compact versions representing logical operations by means of vectorial truth values.

Let us look at the two monadic tables corresponding to logical identity and negation:

p	Identity	Negation
1	1	0
0	0	1

Among the dyadic tables we now show the seven fundamental tables.

	AND	OR	Implication	Equivalence	Exclusive-or	Sheffer's function	Peirce's function
$p\,q$	$p \land q$	$p \lor q$	$p \Rightarrow q$	$p \equiv q$	$p \neq q$	$p \mid q$	$p \downarrow q$
1 1	1	1	1	1	0	0	0
1 0	0	1	0	0	1	1	0
0 1	0	1	1	0	1	1	0
0 0	0	0	1	1	0	1	1

There are also two dyadic operations of particular interest, the Sheffer and Peirce functions.

Since these connectives have relevance in computer science (where they are also called "logic gates"), let us point out their usual names in this context: negation = NOT; conjunction = AND; disjunction = OR; implication = IMPL; exclusive–or = XOR; Sheffer's connective = NAND; Peirce's connective = NOR.

Let us now show with two examples how functions of several variables can be computed from these monadic and dyadic operations.

Example 1:

$$\Phi\,(p,q,r) = (p \vee q) \Rightarrow r$$

then $\Phi\,(0,1,0) = 0$.

Example 2:

$$\Phi\,(p,q,r,s) = [(\neg p \vee q) \vee r] \equiv s$$

then $\Phi\,(0,0,1,1) = 1$.

We point out that here the sign of equality = is an informality that we allow ourselves, but it is not a symbol defined between logical variables (equivalence should be used more rigorously).

Some important consequences emerge from the tables. For example,

$$p \wedge p = p,$$

$$p \vee p = p.$$

In set theory, these propositional operations of a single variable, correspond to the intersection and union of a set with itself:

$$A \cap A = A; A \cup A = A,$$

operations that are valid whether A is the universal set or the empty set.

In current versions of Boolean algebras, where the $\wedge$ is associated with multiplication and $\vee$ with the addition, the previous relationships correspond to the following:

$$1.1 = 1; \quad 0.0 = 0;$$

$$1 + 1 = 1; \quad 0 + 0 = 0.$$

Other interesting formulas are as follows:

$$(p \Rightarrow p) = 1; (p \equiv p) = 1; (p \neq p) = 0; (p \mid p) = \neg p; (p \downarrow p) = \neg p.$$

Let us also note that the table of the conjunction imposes the principle of non-contradiction:

$$(p \wedge \neg p) = 0. \tag{1.7}$$

A property of logical calculus is that with a pair of operations, for example, negation $\neg$ and disjunction $\vee$, the rest of the operations can be expressed. For example,

$$(p \Rightarrow q) = (\neg p \vee q). \tag{1.8}$$

Sheffer's discovery (published in 1913) of his connective had enormous theoretical (and then practical) significance with the advent of electronics, since he showed that with this single operation all other functions could be represented. For example,

$$(p \Rightarrow q) = p|(q|q).$$

Note that Sheffer's function, NAND, is a negation of AND and implicitly merges two operations. The same goes for Peirce's function, NOR.

1.2 The tautologies

The so-called tautologies are formulas $\Phi\left(v_{j(1)}, \ldots, v_{j(r)}\right)$ that are always true whatever the values of the $v_{j(i)}$. Among these tautologies are the "laws" that govern a large part of mathematical demonstrations. We will expose some of the important tautologies using their classic names and closely following the exposition of the logic text by Patrick Suppes (1957). The reader will be able to verify the tautological character of these formulas using the truth tables.

1.2.1 Implicational tautologies

Law of detachment: $p \wedge (p \Rightarrow q) \Rightarrow q$
Modus tollendo tollens: $\neg q \wedge (p \Rightarrow q) \Rightarrow \neg p$
Modus tollendo ponens: $\neg p \wedge (p \vee q) \Rightarrow q$
Modus ponendo ponens: $[p \wedge (p \Rightarrow q)] \Rightarrow q$
Simplification law: $(p \wedge q) \Rightarrow p$
Export law: $[(p \wedge q) \Rightarrow r] \Rightarrow [p \Rightarrow (q \Rightarrow r)]$
Import law: $[p \Rightarrow (q \Rightarrow r)] \Rightarrow [(p \wedge q) \Rightarrow r]$
Law of addition: $p \Rightarrow (p \vee q)$
Law of absurdity: $[p \Rightarrow (q \wedge \neg q)] \Rightarrow \neg p$

1.2.2 Equivalential tautologies

Law of double negative: $p \equiv \neg\neg p$
Law of contraposition: $(p \Rightarrow q) \equiv (\neg q \Rightarrow \neg p)$

De Morgan's law 1: $\neg(p \wedge q) \equiv \neg p \vee \neg q$
De Morgan's law 2: $\neg(p \vee q) \equiv \neg p \wedge \neg q$
Commutative law 1: $p \wedge q \equiv q \wedge p$
Commutative law 2: $p \vee q \equiv q \vee p$
Law of equivalence between implication and disjunction: $(p \Rightarrow q) \equiv \neg p \vee q$
Law of denial of implication: $\neg(p \Rightarrow q) \equiv p \wedge \neg q$
Biconditional law: $(p \equiv q) \equiv (p \Rightarrow q) \wedge (q \Rightarrow p)$
Alternative biconditional law: $(p \equiv q) \equiv (p \wedge q) \vee (\neg p \wedge \neg q)$

1.2.3 Complementary tautologies

Law of the excluded middle: $p \vee \neg p$
Law of contradiction: $\neg(p \wedge \neg p)$

Thus, the law of the excluded middle allows us to confirm as true the following proposition: "The previously stated tautologies are either true or they are not." Also, equivalently: "Previous tautologies are false or they are not."

1.3 Note on George Boole's formalism

We have exposed elsewhere (Mizraji 2013) the personal and technical adventures of Boole. Here we will only briefly show the strange way in which he achieved one of his most important constructions: logical polynomials. In his theory, Boole (1847, 1854) defines the symbols 1 and 0. This symbol 0, which satisfies the usual rules of numeracy arithmetic ($0.a = a.0 = 0$), is transferred to the domain of logic as the empty class. Boole says: "the symbol 0 represents 'Nothing.'" In Boole's class theory, multiplications are what we now call intersections. Regarding 1, he describes its analogy with arithmetic ($1.a = a.1 = a$) and defines it as a class that represents the Universe. His definitions imply that between Nothingness and the Universe are located the various classes that he will have to consider (Boole 1854, chapter III, proposition II: "In fact, Nothing and Universe are the two limits of class extension [. . .]").

Other properties of its classes are as follows:
(i) If x represents one class of objects, $1 - x$ represents the opposite, the complementary class of objects; by definition we know that $x(1 - x) = 0$, which is a version of the principle of non-contradiction.
(ii) Given two classes x and y, $x.y = y.x$ is true.
(iii) For a class x, the idempotency is fulfilled $x^2 = x$, and iterating the multiplication n times results in $x^n = x$.

From these properties arises another proof of the principle of non-contradiction.
Since
$x^2 = x$, it turns out that $x - x^2 = 0$, and from here it follows

$$x(1-x) = 0, \tag{1.9}$$

which is the Boolean version of propositional equivalence (1.7).

This is what Boole saw as the key to his theory: to be able to find under the laws
of algebra laws of logical thought, such as this law of non-contradiction. To be a tree
and a non-tree is impossible, it contravenes our basic logic.

In his theory, Boole performs a bit of magic, and now we will look at one of his
tricks. The goal now is to express any function $f(x)$ of a logical variable. To this end,
Boole proposes a series expansion of powers according to MacLaurin's formula, no
doubt aware that the derivatives of this expansion do not follow the conditions that
mathematical rigor demands. It turns out that

$$f(x) = f(0) + f'(0)x + \frac{f''(0)}{1.2}x^2 + \cdots . \tag{1.10}$$

But since all the powers of x are equal to x, it follows:

$$f(x) = f(0) + \left[f'(0) + \frac{f''(0)}{1.2} + \cdots \right] x. \tag{1.11}$$

Since this expression also applies to $x = 1$, it is

$$f(1) = f(0) + \left[f'(0) + \frac{f''(0)}{1.2} + \cdots \right]. \tag{1.12}$$

Therefore, the parentheses containing the derivatives can be expressed as

$$f(1) - f(0) = \left[f'(0) + \frac{f''(0)}{1.2} + \cdots \right]. \tag{1.13}$$

This allows us to introduce the eq. (1,13) into (1,11), thereby eliminating the derivatives. Hence,

$$f(x) = f(0) + [f(1) - f(0)]x, \tag{1.14}$$

and from here

$$f(x) = f(1)x + f(0)(1-x). \tag{1.15}$$

This is the first logical polynomial calculated by Boole. These polynomials are extended for the number of variables needed. For two variables x and y, their structure is

$$f(x.y) = f(1,1)xy + f(1,0)x(1-y) + f(0,1)(1-x)y + f(0,0)(1-x)(1-y). \quad (1.16)$$

But the magic continues through the apparently arbitrary rules that Boole stipulates for the coefficients of these logical polynomials. These rules are as follows: For functions of logical variables to which these polynomial developments are applied, if the calculated coefficients (e.g., $f(1,1)$ and $f(1,0)$) are different from 1 or 0 (calculated according to the rules of standard number arithmetic where $0/1 = 0$), then the factors associated with those coefficients are 0. If, on the contrary, the coefficients are 1, the factors subsist and if they are 0, the factors are cancelled out by the rule of the product by 0. In the event of a situation of $0/0$ indeterminacy, Boole proposes a solution that does not necessarily cancel out the associated factor. However, for $1/0$ the expression it affects is zero. The term of the development is also null if its coefficient is greater than 1. We will see all this in action through an important example.

Boole created an arbitrary variable v which means "some." So if z represents the class of mortal beings and if x represents humans, then

$$x = vz. \quad (1.17)$$

It means that humans are some of the mortals. Let us now try to analyze the logical meaning of v, and move to standard arithmetic by writing

$$v = \frac{x}{z} = f(x,z). \quad (1.18)$$

Let us apply for $f(x,z)$ the development of eq. (1.16):

$$f(x,z) = \frac{1}{1}xz + \frac{1}{0}x(1-z) + \frac{0}{1}(1-x)z + \frac{0}{0}(1-x)(1-z).$$

Defining $0/0$ as a variable v^* that means "maybe some," two conclusions emerge from this polynomial:

$$(a)\ v = xz + v^*(1-x)(1-z),$$

$$(b)\ x(1-z) = 0.$$

Returning to Boolean logic, (a) shows that v has a first term that is a logical product between humans and mortals. The second term of (a) indicates that "maybe some" non-humans are immortal, which allows for the existence of deities and avoids possible censure by some clergy. Conclusion (b) points out that there are no immortal humans, another interesting conclusion, which, in a way, validates Boole's heuristic regarding the way to interpret the coefficients of their polynomials.

Let us look at one last example. We will try to show that the addition of non-null Boolean variables in the form that it is adopted today $(1+1=1)$ is not the one that Boole defined. Let us write a Boolean addition as a function of two variables:

$$x + z = f(x, z)$$

and let us apply polynomial development:

$$f(x, z) = 2xz + 1x(1 - z) + 1(1 - x)z + 0(1 - x)(1 - z).$$

The conclusions are:

$$(a)\ 2xz = 0$$

$$(b)\ f(x, z) = x - xz + z - zx = x + z.$$

This procedure involves fully accepting the validity of the polynomial expansion (1.16) for any pair of logical variables, even when, as in the last example, both coincide. Now if we apply this expansion to the sum $x + x = f(x, x)$ result, then

$$f(x, x) = x(1 - x) + (1 - x)x.$$

But each of the terms is a representation of the principle of non-contradiction (1.9) and is therefore null. Therefore, if $x + x = 0$, it turns out then that the original laws of the Boolean addition are: $1 + 1 = 0$, $1 + 0 = 1$, $0 + 1 = 1$, $0 + 0 = 0$. Hence, in this algebra, addition is the or-exclusive of propositional calculus. Jevons' modification, making $1 + 1 = 1$, corresponds to the disjunction of the truth tables.

Is Boolean algebra a binary algebra? Maybe not, since the classes in which its logic is concentrated (the set of humans, etc.) are generally subsets of the universal set (mortal beings). The propositional logic represented in the tables in the previous section arises from restricting Boolean operations to the universal and empty sets, and from mapping these sets to the veritative values "true" = 1 and "false" = 0. This is clearly seen if the coefficients of the polynomials of a variable are replaced by the four possible sets that are formed from 0 and 1. In this way, the four monadic functions of the propositional calculus are obtained. At the same time, if we also operate with the polynomials of two variables, it arises that, when the variables are replaced by 0 and 1, we rediscover the 16 functions of the propositional calculus.

References

Boole, G. (1847, 1984) The Mathematical Analysis of Logic, Chair, Madrid.
Boole, G. (1854, 1958) An Investigation of the Laws of Thought, Dover, New York.
Lukasiewicz, J. (1934) On the history of the logic of prepositions, reprinted in J. Lukasiewicz, Selected
    Works, pp. 197–217 (L. Borkowski, Ed), North-Holland, 1980, Amsterdam.
Mizraji, E. (2013) En Busca de las Leyes del Pensamiento (Second Edition), Trilce-Dirac, Montevideo.
Suppes, P. (1957) Introduction to Logic, Van Nostrand, New York.
Wittgenstein, L. (1921) Tractatus-Logico-Philosophicus, Routledge, London.

Chapter 2
An informal look at matrices and vectors

The search for general procedures for solving systems of linear equations led to the creation of determinants and this paved the way for the invention of matrices. The official date of birth of matrices is December 10, 1857, and its dissemination occurred on January 14, 1858, when Arthur Cayley read before the Royal Society of London "A Memoir on the Theory of Matrices," a text later published in the *Philosophical Transactions of the Royal Society*.

Matrix theory remained semi-hidden in the rich universe of algebra at the end of the nineteenth century until the historic events of 1925. That year, Werner Heisenberg found a strange novel representation for the variables observable in quantum mechanics. This representation involved arranging the variables in infinite-dimensional tables, and these variables, for physical reasons, had to be governed by precise mathematical operations that were unknown to Heisenberg and many of his colleagues. Finally, Max Born, one of Heisenberg's mentors, in one of his moments of relaxation recalled the distant algebra classes he received in Breslau from his teacher Jakob Rosanes. There Rosanes had taught his students notions about algebraic objects called matrices, with particular multiplication rules and which they usually did not commute. And there Born realized that this was what Heisenberg had found: matrices. This chronicle can be consulted at the Nobel Lecture in Born (1954).

After this heroic breakthrough in physics, matrices spread to extensive territories of the natural sciences, from the dynamics of biological populations or the stability of social systems to research into the control systems of physiology or the neural networks of the brain. At the same time, entire domains of mathematics and statistics were restated through the compact formalism of matrix theory. Recently, the development of powerful computing capabilities has led to a renaissance of matrix theory. This happened because problems and procedures emerged that required operations involving thousands or hundreds of thousands of variables, such as in the domains of image processing (Jain 1989) or data mining techniques (Berry and Browne 2005). And now large language models (LLMs) and generative artificial intelligence have exacerbated the need to compute very large dimensional matrices.

2.1 The matrices

In what follows we will base part of the initial definitions and examples on the text of Hohn (1964). A matrix is an array of numbers or functions consisting of m rows and n columns. It can be presented as follows:

https://doi.org/10.1515/9783112230053-003

$$A = \begin{bmatrix} a_{11} & \cdots & a_{1n} \\ \vdots & \ddots & \vdots \\ a_{m1} & \cdots & a_{mn} \end{bmatrix}. \qquad (2.1)$$

A shorthand way to represent this matrix is as follows:

$$A = [a_{ij}]_{mn}, \qquad (2.2)$$

and we will say that it is a matrix of order $m \times n$. A matrix for which the number of rows and columns ($m = n$) coincide is called a square matrix. When we refer to square matrices we will simplify the notation (2.2), $A = [a_{ij}]$, and we can talk about matrices of order m.

The first property to point out is the equality of matrices. If $A = [a_{ij}]_{mn}$ and $B = [b_{ij}]_{mn}$, we will say that $A = B$ when for every pair ij is $a_{ij} = b_{ij}$. Another important property is transposition. We define the transpose of matrix A as

$$A^T = [a_{ji}]_{nm}.$$

We show the transposition in the following example:

$$A = \begin{bmatrix} 1 & 4 \\ -2 & 0 \\ 1 & -3 \end{bmatrix}, A^T = \begin{bmatrix} 1 & -2 & 1 \\ 4 & 0 & -3 \end{bmatrix}.$$

A square matrix is symmetrical if $A = A^T$.

The addition of matrices is only possible between matrices of the same order. If $A = [a_{ij}]_{mn}$ and $B = [b_{ij}]_{mn}$, then the matrix $C = A + B$ is defined as $C = [c_{ij}]_{mn}$ with

$$c_{ij} = a_{ij} + b_{ij}. \qquad (2.3)$$

Note that addition is commutative: $A + B = B + A$, and it is easy to prove that it is also associative. A null matrix is definable for any order. For mn there is a matrix $0 = [0_{ij}]_{mn}$, that is, a neutral for addition: $A + 0 = 0 + A = A$.

Matrices can be multiplied by scalars according to the following rule: if α is a scalar, then $\alpha A = \alpha[a_{ij}]_{mn} = [\alpha a_{ij}]_{mn}$. If you choose the scalar -1, the preceding rule allows you to define for any A its opposite matrix $-A$. And from here, it turns out that $A + (-A) = A - A = 0$.

Square matrices have additional properties. Given a square matrix of order m

$$A = [a_{ij}],$$

the set of elements $a_{ii}, i = 1, \ldots, m$, is called the main diagonal of A, and its sum is called the trace of A, and it is annotated $\operatorname{tr} A = a_{11} + \cdots + a_{mm}$.

For square matrices, you can define a unit matrix, I, that has only 1's on its main diagonal and 0's in all other positions. It can be compactly defined as

$$I = [\delta_{ij}],$$

where δ_{ij} is the Kronecker delta ($\delta_{ij} = 0$ if $i \neq j$; $\delta_{ij} = 1$ if $i = j$).

It is in multiplication that matrices show their most novel properties. Not every pair of matrices can be multiplied. For this operation to be possible, it is required that the number of columns of the first factor matches the number of rows of the second factor. That is to say that A_{mn} can be multiplied by B_{np}, resulting in

$$C_{mp} = A_{mn}\ B_{np}. \tag{2.4}$$

In this case, it is said that A *pre multiplies* B and that B *post multiplies* A. Note that in this example the product BA does not exist. When two matrices are capable of being multiplied, they are usually called *"conformable"* for multiplication.

If $A = [a_{ij}]_{mn}$, $B = [b_{ij}]_{np}$, and $A.B = C = [c_{ij}]_{mp}$, then each component of C is defined by the following multiplication rule:

$$c_{ij} = a_{i1}b_{1j} + a_{i2}b_{2j} + \cdots + a_{in}b_{nj} = \sum_{k=1}^{n} a_{ik}b_{kj}. \tag{2.5}$$

Let us exemplify this with two numerical matrices:

$$\begin{bmatrix} 1 & -1 & 2 \\ 3 & 0 & 1 \end{bmatrix}_{2\times3} \begin{bmatrix} 1 & 2 & 0 \\ 0 & -1 & 1 \\ 1 & 2 & -1 \end{bmatrix}_{3\times3} = \begin{bmatrix} 3 & 7 & -3 \\ 4 & 8 & -1 \end{bmatrix}_{2\times3}.$$

The matrix product is associative, so that $(AB)C = A(BC)$.

Regarding commutativity, let us note, to begin with, that in the case of the previous example, multiplication is not commutative because these matrices are not conformable for commutation. On the other hand, if you have a matrix A_{mn} and another B_{nm}, both products are possible, $A_{mn}B_{nm} = C_{mm}$ and $B_{nm}A_{mn} = D_{nn}$, but these matrices C and D are necessarily different because their dimensions differ, which is another way of not commuting. We will illustrate the situation of the square matrix product through a numerical example. For two matrices A and B of order m, we will have two possible products $AB = C$ and $BA = D$, with C and D also of order m. Let us look at the example:

$$\begin{bmatrix} 2 & -1 \\ -1 & 2 \end{bmatrix} \begin{bmatrix} 1 & 4 \\ -1 & 1 \end{bmatrix} = \begin{bmatrix} 3 & 7 \\ -3 & -2 \end{bmatrix},$$

$$\begin{bmatrix} 1 & 4 \\ -1 & 1 \end{bmatrix} \begin{bmatrix} 2 & -1 \\ -1 & 2 \end{bmatrix} = \begin{bmatrix} -2 & 7 \\ -3 & 3 \end{bmatrix}.$$

In general, it also does not commute the product of square matrices. But in some particular cases, it can be commuted. The easiest to show occurs with the identity matrix. For a matrix A of order m and the identity matrix I of order m, the following is true:

$$AI = IA = A.$$

Each square matrix can be multiplied by itself, obtaining successive powers of the same order:

A, $AA = A^2$, $AAA = A^3$, etc.

Note in passing that commutative matrices arise naturally, because if we define $B = A^2$ then $AB = BA$. There are commutative matrices of more subtle structure, but we will not deal with this topic here. Let us point out that the identity matrix is always idempotent: $I^2 = I$.

In the field of real numbers, basic properties such as commutativity ($ab = ba$) are true, but the cancelling property of the product is also true (if $a \neq 0$, $ab = ac \Rightarrow b = c$) and the non-existence of zero divisors ($ab = 0 \Rightarrow a = 0$ or $b = 0$). Matrices generally may not comply with the cancelling property of the product and may have divisors of zero. Two examples taken from Hohn's text are shown:

We first show three matrices A, B, and C, where $AB = AC$ with $B \neq C$:

$$A = \begin{bmatrix} 1 & 2 & 0 \\ 1 & 1 & 0 \\ -1 & 4 & 0 \end{bmatrix}; \quad B = \begin{bmatrix} 1 & 2 & 3 \\ 1 & 1 & -1 \\ 2 & 2 & 2 \end{bmatrix}; \quad C = \begin{bmatrix} 1 & 2 & 3 \\ 1 & 1 & -1 \\ 1 & 1 & 1 \end{bmatrix}.$$

Then

$$AB = AC = \begin{bmatrix} 3 & 4 & 1 \\ 2 & 3 & 2 \\ 3 & 2 & -7 \end{bmatrix}.$$

And now we show a case of the existence of zero divisors, where $AB = 0$, $A \neq 0$ y $B \neq 0$:

$$\begin{bmatrix} 1 & 2 & 0 \\ 1 & 1 & 0 \\ -1 & 4 & 0 \end{bmatrix} \begin{bmatrix} 0 & 0 & 0 \\ 0 & 0 & 0 \\ 1 & 4 & 9 \end{bmatrix} = \begin{bmatrix} 0 & 0 & 0 \\ 0 & 0 & 0 \\ 0 & 0 & 0 \end{bmatrix}.$$

A relevant property of square matrices is that there are matrices that have an inverse matrix. In that case, for A of order m, there is a matrix B of order m (which is unique) and that it fulfills:

$$AB = BA = I.$$

This is noted $B = A^{-1}$. Let us point out that the "pathological" cases we have just shown, extracted from Hohn's text, occur when matrices do not have an inverse.

The relevance of the existence of inverse arises from the fact that in this situation a matrix equation can be formally solved even if it involves hundreds or thousands of variables. Given the matrix equation

$$AX = B, \tag{2.6}$$

where matrix X contains the unknowns, if A has an inverse, then the solution to this equation is

$$X = A^{-1} B. \tag{2.7}$$

Each square matrix A is associated with a number called a determinant $\det(A)$. This number is easy to calculate for matrices of order 2 and 3, and more complicated in higher orders, but always calculable through its definition. We will dispense here with defining the determinants because it is a complicated topic, and we refer to the standard texts of linear algebra. It is nevertheless relevant to mention them because they define the existence of the inverse of a square matrix. Invertible matrices are called non-singular. The crucial theorem here, which we will only mention, is this: *The necessary and sufficient condition for the inverse of a matrix to exist is that its determinant is not zero.*

A fundamental property of square matrices is the existence of a set of scalars called eigenvalues. These eigenvalues are defined by the equation

$$Ax = \lambda x, \tag{2.8}$$

where λ is a scalar. Vectors associated with eigenvalues are called eigenvectors. The equation that allows the calculation of eigenvalues is

$$\det(A - \lambda I) = 0. \tag{2.9}$$

This determinant generates polynomials in λ of degree equal to the order of the matrix. The set of λ, the roots of eq. (2.9), is called the "spectrum" of matrix A.

To end this section, let us ask ourselves the question: are only square matrices invertible? For any matrix, there is another associated matrix called the Moore-Penrose pseudoinverse (or generalized inverse). The pseudoinverse of matrix A, annotated as A^+, is defined by a series of axioms (see Barnett (1990) for this and other generalized inverses):

(i) $AA^+A = A$,
(ii) $A^+AA^+ = A^+$,
(iii) $(AA^+)^T = AA^+$,
(iv) $(A^+A)^T = A^+A$.

By developing this formalism, it can be shown that a matrix equation involving rectangular matrices

$$AX = B,$$

does not usually have an exact solution, but has an approximate solution given by

$$X \cong A^+ B. \tag{2.10}$$

This solution is, in general, an optimal approximation according to the criteria of proximity by least squares to what would be the virtual exact solution. However, in the case where the columns of matrix A are linearly independent, there may be an exact solution and also an explicit expression for the pseudoinverse given by

$$A^+ = (A^T A)^{-1} A^T. \tag{2.11}$$

2.2 Vectors

An n-dimensional vector is a rectangular matrix $[a_{ij}]_{1n}$ or $[a_{ij}]_{n1}$. The first case is called a row vector (1 row and n columns) and the second column vector. Throughout this text we will adopt column vectors as the basis of the representations.

Given a matrix A_{mn} and an n-dimensional column vector x, in the equation

$$Ax = y,$$

y is an m-dimensional column vector. For typographic convenience, an n-dimensional column vector can be written like this:

$$x = [x_1 \ x_2 \ \cdots \ x_n]^T.$$

Given two vectors x and y of order n (or of dimension n) whose components are real numbers, the scalar product is defined as

$$\langle x, y \rangle = x_1 y_1 + \cdots + x_n y_n = \sum_{i=1}^{n} x_i y_i. \tag{2.12}$$

Because they are two column vectors, this definition can also be expressed as a matrix operation:

$$\langle x, y \rangle = x^T y. \tag{2.13}$$

Notice a subtle point: the equality imposed in eq. (2.13) assumes a complete isomorphism between the real numbers and the matrices of order 1×1. So, we will assume this isomorphism without further comment.

Given an n-dimensional vector x, its modulus is defined as

$$|x| = \sqrt{\langle x, x \rangle}. \tag{2.14}$$

Consequently,

$$|x| = \sqrt{x_1^2 + x_2^2 + \cdots + x_n^2}.$$

which is an extension of the Pythagorean theorem. Any vector x can have a modification of its modulus without altering its direction. In particular, modulus 1 vectors, also called normal, are important. The normal vector associated with x is

$$\hat{x} = \frac{x}{|x|}.$$

Two vectors x and y are orthogonal if $\langle x, y \rangle = 0$ y are parallel if $\langle \hat{x}, \hat{y} \rangle = 1$. Orthonormal vectors are both normal and orthogonal.

Let us conclude by pointing out that the angle between two n-dimensional vectors is measured by their cosine, which is given by

$$\cos(x, y) = \frac{\langle x, y \rangle}{|x|.|y|}. \tag{2.15}$$

This cosine is, in fact, the way to define the statistical correlation between two collections of measures.

2.3 Partitioned matrices and Kronecker products

A matrix can be written by grouping its components into sub-matrices. A matrix written in this form is called a partitioned matrix. Partitioning a matrix can have both symbolic and computational advantages. Let us exemplify a partition. Let it be A_{mn} and let us write it like this:

$$A_{mn} = \begin{bmatrix} U_{ab} & V_{cd} \\ W_{ef} & X_{gh} \end{bmatrix},$$

where $a + e = c + g = m$ and $b + d = f + h = n$. The U, V, W, and X submatrices pack the A elements corresponding to the areas they cover, keeping them a_{ij} in their original positions. The transposed matrix of a partitioned matrix like the preceding one is given by the following equation:

$$A^T = \begin{bmatrix} U^T & W^T \\ V^T & X^T \end{bmatrix}.$$

Suppose we partition a matrix and a vector so that the products between submatrices (including subvectors) are conformable. Therefore, it turns out that

$$\begin{bmatrix} U & V \\ W & X \end{bmatrix} \begin{bmatrix} E \\ F \end{bmatrix} = \begin{bmatrix} UE + VF \\ WE + XF \end{bmatrix}.$$

If you partition a matrix based on its columns or rows, you will find interesting results. Given two conformable matrices for product A_{mn} and B_{np}, we can write A as a column m of n-dimensional row vectors:

$$A = \begin{bmatrix} a_1 \\ \vdots \\ a_m \end{bmatrix}.$$

B can be written as a row of n-dimensional column vectors:

$$B = \begin{bmatrix} b_1 \cdots b_p \end{bmatrix}.$$

Note that the product AB can be expressed by scalar products

$$AB = \begin{bmatrix} a_1 \\ \vdots \\ a_m \end{bmatrix} \begin{bmatrix} b_1 \cdots b_p \end{bmatrix} = \begin{bmatrix} \langle a_1, b_1 \rangle & \cdots & \langle a_1, b_p \rangle \\ \vdots & \ddots & \vdots \\ \langle a_m, b_1 \rangle & \cdots & \langle a_m, b_p \rangle \end{bmatrix}.$$

These scalar products arise because we are matrix multiplying an n-dimensional row vector by an n-dimensional column vector. Thus, $[a_h]_{1 \times n}[b_k]_{n \times 1} = (\langle a_h, b_k \rangle)_{1 \times 1}$.

We now show an interesting property of partitioned matrices that is useful for understanding an algebraic problem that emerged during the investigation of the nature of associative memories (a topic that we will briefly discuss in the next chapter). Let there be a set of operations $Mf_1, Mf_2, \ldots, Mf_K$, where the f_i are column vectors and M is a matrix. If the following partitioned matrix is constructed

$$F = [f_1 \ f_2 \ \cdots \ f_K],$$

then it can be shown that in the MF operation, the matrix M behaves similarly to a scalar multiplied by a vector, resulting in

$$MF = [Mf_1 \ Mf_2 \ \cdots \ Mf_K]. \tag{2.16}$$

We conclude this chapter by presenting Kronecker's product, an operation of central importance to the matrix formalism of logic that we will see in the following chapters. Kronecker's abstract definition of the product is based on expressing the product matrix in a partitioned way (Graham 1981; Jain 1989). Given A_{mn} and B_{pq} Kronecker's product is defined as

$$A_{m \times n} \otimes B_{p \times q} = \begin{bmatrix} a_{ij} B \end{bmatrix}_{mp \times nq}. \tag{2.17}$$

This product always exists, whatever the dimensions of the matrices. It is also clear that, in general, it is not commutative. Let us look at a small example for square matrices of order 2:

$$A = \begin{bmatrix} 1 & -3 \\ 4 & 0 \end{bmatrix} ; \qquad B = \begin{bmatrix} 2 & 0 \\ 3 & -1 \end{bmatrix} ;$$

$$A \otimes B = \begin{bmatrix} 1.B & -3.B \\ 4.B & 0.B \end{bmatrix} = \begin{bmatrix} 2 & 0 & -6 & 0 \\ 3 & -1 & -9 & 3 \\ 8 & 0 & 0 & 0 \\ 12 & -4 & 0 & 0 \end{bmatrix} .$$

The basic properties of the Kronecker product are as follows:

1) $(A \otimes B)^T = A^T \otimes B^T$,
2) $(A \otimes B) \otimes C = A \otimes (B \otimes C)$,
3) $A \otimes (B + C) = A \otimes B + A \otimes C$,
4) $(A \otimes B)(C \otimes D) = (AC) \otimes (BD)$.

Property 4 requires that the matrices be conformable for the matrix products involved.

An important case for this work occurs when properties 1 and 4 are applied to the following Kronecker product, where a, b, c, and d are column vectors, with a and c of dimension p, and b and d of dimension q (eventually it can be $p = q$):

$$(a \otimes b)^T (c \otimes d) = (a^T \otimes b^T)(c \otimes d) = (a^T c) \otimes (b^T d) = \langle a, c \rangle \langle b, d \rangle. \tag{2.18}$$

Finally, let us recursively define Kronecker's successive powers as follows:

$$A^{[1]} = A, A^{[k+1]} = A^{[k]} \otimes A.$$

References

Barnett, S. (1990) Matrices. Methods and Applications, Clarendon Press, Oxford.
Berry, M.W., Browne, M. (2005) Understanding Search Engines, SIAM, Philadelphia.
Born, M. (1954) The Statistical Interpretation of Quantum Mechanics, Nobel Lecture.
Graham, A. (1981) Kronecker Products and Matrix Calculus with Applications, Ellis Horwood, Chichester.
Hohn, F.E. (1964) Elementary Matrix Algebra, Macmillan, New York.
Jain, A.K. (1989) Fundamentals of Digital Image Processing, Prentice Hall, New York.

Chapter 3
Neural memory models and the origin of vector logic

A severely challenging problem for neurobiological research is to devise procedures for storing information in the brains of complex animals, such as humans, that are robust to the physical deterioration of the neural support of memories. This problem arose from evidence from both neurological pathology and animal experimentation, which found that, in many situations, the information stored in the memories persisted without extreme deterioration despite the existence of significant neurological lesions.

Toward the end of the 1960s and the beginning of the 1970s, several researchers independently converged toward a mathematical formalism that suggests a possible solution to this reliability problem. The biological basis of this solution was the fact that, inside the brain, information is encoded by a set of electrochemical signals carried by the axons of neurons. For example, a colored image with a certain texture captured by our retinas penetrates the brain carried by hundreds of thousands of electrochemical signals that travel in the axons of the optic nerves. This set of signals, after several instances of preprocessing in different neural modules, finally reaches a region of the brain where the cognitive analysis of the signals and the recognition of the nature and properties of the image occur.

These sets of signals are assumed to be present in all instances associated with cognition, either in the analysis of sensory information, as in the previous example, or in the development of the cognitive activities that define psychism. To represent these sets of signals carried by hundreds or hundreds of thousands of axons in parallel, the representation of these collections of signals by vectors (possibly vectors of hundreds of thousands of components) was adopted as a formalism. Let us now look at the way in which the Finnish researcher Teuvo Kohonen (1977) posed the problem of defining the structure of a neural memory based on this vector representation.

Kohonen defines an associative memory (which, e.g., associates faces with corresponding names) as a set of ordered pairs of column vectors

$$\text{Mem} = \left\{ (g_1, f_1), (g_2, f_2), \ldots, (g_K, f_K) \right\}, \tag{3.1}$$

where in each pair the f_i are n-dimensional vectors that encode the faces, and the g_i are m-dimensional vectors that encode the names associated with the f_i corresponding ones. In this case, f_i is the stimulus vector of memory and g_i is the associated response.

Kohonen's problem is this: Is there a matrix M such that when it receives f_i generates its corresponding vector g_i for $(g_i, f_i) \in \text{Mem}$? This author transforms this ques-

https://doi.org/10.1515/9783112230053-004

tion into the next algebra problem. Let us construct the following partitioned matrices:

$$F = [f_1 \ f_2 \ \cdots \ f_K]; \quad G = [g_1 \ g_2 \ \cdots \ g_K].$$

The math problem now is to solve M in the following equation:

$$G = MF \tag{3.2}$$

(see eq. (2.16)). A formal and approximate solution arises from the Moore-Penrose pseudoinverse:

$$M \cong GF^+. \tag{3.3}$$

If the vectors of F are linearly independent, an exact solution results:

$$M = G(F^T F)^{-1} F^T. \tag{3.4}$$

And if the vectors of F are orthonormal (orthogonal of modulus 1), an explicit solution is obtained:

$$M = \sum_{i=1}^{K} g_i f_i^T. \tag{3.5}$$

It should be noted that orthogonality can be justified as an approximation if we are dealing with vectors of very high dimension and the data is very diluted, that is, with a preponderance of zeros among their components. Normalization is a simple mathematical transformation and does not modify the angles, which, in this theory, play the fundamental role in pattern recognition.

This simple expression (3.5), despite the many approximations and simplifications involved, was key to seeing that these memories had very remarkable properties, which, among other things, suggested a solution to the problem of robustness in the case of physical deterioration of the support (for details, see Kohonen 1977; Mizraji 2008). This is because this equation shows that the associated vector pairs are scattered among all the coefficients in the matrix. And if the matrix is very large, with millions of components, as is expected in a module that supports a neural memory, the destruction of a few hundred or thousands of its components only causes the loss of small fragments of the stored information. In the translation "model → biology," the vectors carry the electrochemical codes and the coefficients of the matrices represent the properties of the synapses ("synaptic weights" in the jargon of the subject).

The way these matrices operate is also clearly shown in the following equation:

$$Mf_h = \sum_{i=1}^{K} g_i \langle f_i, f_h \rangle. \tag{3.6}$$

Given the orthonormality, it happens that if f_h does not belong to some pair of the Mem memory, then

$$Mf_h = 0.$$

But if it belongs to memory, it is

$$Mf_h = g_h,$$

and the required association occurs. Note that, in this memory, the identification depends on a scalar product, which means that the similarity between vectors is measured by the angle between them.

This theory that we are schematizing has many rich derivations that we must omit here. Also, in the 1970s, physicist Leon N. Cooper, based on how these memories could be generated in the brain, suggested that concepts could be identified with the vectors involved in these matrix memories. The reason for this suggestion is that in the process of memory generation, the final vectors are prototypes, in fact averages (in the mathematical and not metaphorical sense), arising from the various vectors with which memory was initially stimulated (Cooper 1973).

A panoramic review of how neural models evolved into modern LLMs and generative AI can be found in Valle-Lisboa et al. (2023).

3.1 Memories, contexts, and logic: a personal note

I will now pass temporarily to the first person because I now need to describe an experience of my own. I became acquainted with the matrix models of distributed memories at the end of 1976. Along with the fascination they gave me, a problem arose, inherited from my own history as a student.

Very early in my university life I had known two books by William Ross Ashby (1952, 1956). From them I got a notion that I could never abandon: a living being is necessarily an adaptive machine. Its survival is based on the fact that it manages to maintain certain essential variables (in our case, the concentrations of glucose and oxygen in the arterial blood and the pH of the internal environment, among others) within a narrow range of variability. To achieve this, it must have the ability to change its "mechanical" modes of operation if environmental circumstances change. The modes of operation are parameterized by the context in which the events occur. These contexts modulate behaviors (in humans both physiological and behavioral). All this machinery is contingent and is created by natural selection: the forms of behavior and their contexts are adaptations to our probable environments. If we fall into a lake of lava, there is no physiological control possible for our survival.

In short, our adaptive machinery associates states $e_i \in E$ and contexts $p_j \in P$ through functions

$$\Phi : P \times E \to E,$$

where you move from one state to another under the rule that stipulates a contextual parameter:

$$e_i(j,\ t+1) = \Phi \left[p_j(t),\ e_i(t) \right].$$

Second, I was influenced by Jacques Monod's ideas on allosteric control (Monod 1972). Allosteric control is evidenced in the genetic and biochemical machinery of cells, where there are certain enzymes or receptors that, together with their natural ligands, are sensitive to molecular signals that modulate their action. This determines that multiple biochemical and genetic events have the possibility of adapting to conditions imposed by their natural environments. These ideas of Monod's, experimentally supported, seemed to me a natural translation of Ashby's notions to the domain of molecular biology. Moreover, these findings from molecular biology provide a solid factual basis for explaining the compatibility of Darwinian evolution with the principles of physics. To this, Monod adds another relevant notion to support the physical consistency of biological evolution: gratuitousness.

"Gratuity" in this framework means that the signals that act as modulators need not have chemical relationship or similarity to the basic ligands of the enzymes or allosteric receptors. This gratuitousness is at the basis of evolutionary freedom to adapt to new environments. It is therefore natural to relate this idea about allosteric regulators to the parameters that change the modes of behavior in Ashby's adaptive machines (for an in-depth analysis of these issues, see Mizraji (1999)).

The problem I alluded to at the beginning of the chapter (neural reliability), framed within the ideas of Ashby and Monod, led me to the following problem: How to preserve the magnificent properties of matrix memories, but modulating these memories using vectors that carry contextual information? And also that these vector contexts could be completely arbitrary in relation to the input–output vector pairs stored in memory. This was a neurobiological version of gratuitousness. The natural motivation for this problem is that the central nervous system of a human being is his most refined system of adaptive control, where the impositions of Darwinian evolution are joined by the ability to learn innumerable modes of behavior. These include things as varied as systems of adaptation to environments (e.g., clothing), articulate language, logical reasoning, the development of science and medicine, the technologies associated with those sciences, ships, automobiles, airplanes, and a very long etcetera. All these creations are consequences of the adaptive abilities of the brain modules that make up our cognitive life.

To focus more on the problem, I usually use this situation: If we see an image of a dog and they ask us what the animal's name is, in a Hispanic environment we can answer "Perro," but if we are in an Anglo-Saxon environment and we know some English, we may be able to answer "Dog." In other words, the same input (image of the dog) produces two different responses depending on the idiomatic context in which we are immersed. This is the aspect associated with adaptation. But the requirement of gratuitousness arises from the fact that there is no neurological obstacle for humans in the year 2115 to call the current dog (given the natural evolution of languages and their lexicon), "Ladral" or "F43 Mammal," words that their nervous systems would associate with the image of the dog that entered through their optic nerves.

How to adapt a matrix memory to achieve this? A possible solution is to add the contexts to the input vectors:

$$e_i^T(j) = [f_i^T \mid p_j^T],$$

where the input f_i is expanded by a context so that the input vector sums the dimensions of both vectors. But the linearity of the memory matrix prevents this procedure from being used if we are looking for gratuitness for the context. On other approach, the contextualization with matrix formalism is abandoned and networks with hidden layers and nonlinear links are used (for details, see Anderson (1995)).

The procedure I found in 1987 to preserve the matrix structure of the memories was to combine the contexts and the inputs using Kronecker's product (Mizraji 1989). With this formalism, the model of maximum simplicity (as in eq. (3.5)) has this structure:

$$M = \sum_{j,i} g_{ji}(p_j \otimes f_i)^T. \tag{3.7}$$

Here, the same f with two different contexts p produces different associations g. When M receives a contextualized input, the following occurs:

$$M(p_b \otimes f_a) = \sum_{j,i} g_{ji}\langle p_j, p_b \rangle \langle f_i, f_a \rangle. \tag{3.8}$$

This double filter, created by the two scalar products, is the basis for different contexts and the same input to produce different associations. This type of neural configuration that modulates associations through multiplicative contexts has proven useful for representing the grammatical structures of human language (beim Graben et al. 2008; beim Graben and Gerth 2012). In neural reality, nothing as perfect as a Kronecker product can exist, but model (3.7), slightly deteriorated in terms of the structure of its matrix, retains a good part of its properties if the dimensions of the vectors are large (Pomi and Mizraji 1999).

For several of the basic neural models, such as logical networks (McCulloch and Pitts 1943), the perceptron (Rosenblatt 1958), and matrix memories (Kohonen 1977; Anderson 1995), the computation of the XOR logic function has presented various difficulties (surmountable or not). Therefore, it was natural to evaluate how this model of memory with multiplicative contexts performed against the XOR computation. In the short paper of 1989 in which I presented this model, there is a section of a few lines entitled "Quasi-Logical Behaviour" and there this matrix is described:

$$X = s\ (a\otimes b)^T + s\ (b\otimes a)^T + n\ (a\otimes a)^T + n(b\otimes b)^T, \tag{3.9}$$

where s and n are vectors representing, respectively, "true" and "false," and a and b are orthonormal vector patterns representing opposite concepts. This memory processes its inputs as follows:

$$X(a\otimes b) = X(b\otimes a) = s,$$
$$X(a\otimes a) = X(b\otimes b) = n.$$

Cautiously, the article points out that this behavior is similar to binary logical networks that execute the XOR logic function. Although the article later takes other directions, this section aroused my interest in adapting the formalism of matrices and vectors to the basic operations of elementary logical calculus. Thus, varied results emerged (Mizraji 1994) that were covered by the name "vector logic" due to the vector nature of veritative values.

Note: Mario H. Otero, a distinguished professor of logic and history of mathematics at our university, was a reviewer of the work I published in Galileo in 1994. To the conversations I had with him during this process and to his teachings, I owe the knowledge of the remarkable impact of nineteenth-century British algebraists, such as Babbage, Peacock, Gregory, De Morgan, Cayley, and Boole, in the development of symbolic methods inspired by Leibniz's formalisms, methods that later enriched innumerable areas of scientific research. This knowledge that I obtained from Mario Otero was a fundamental guide for my later work.

References

Anderson, J.A. (1995) An Introduction to Neural Networks, MIT Press, Cambridge.

Ashby, W.R. (1952) Design for a Brain, Chapman and Hall, London.

Ashby, W.R. (1956) An Introduction to Cybernetics, Wiley, New York.

Beim Graben, P., Pinotsis, D., Saddy, D., Potthast, R. (2008) Language processing with dynamic fields, Cognitive Neurodynamics, 2: 79–88.

Beim Graben, P., Gerth, S. (2012) Geometric representations of minimalist grammars, Journal of Logic, Language and Information, 21: 393–432.

Cooper, L.N. (1973) A possible organization of animal memory and learning, in Proceedings of the Nobel Symposium on Collective Properties of Physical Systems, Aspen Garden, Sweden.

Kohonen, T. (1977) Associative Memory: A System-Theoretical Approach, Springer, New York.

McCulloch, W.S., Pitts, W. (1943) A logical calculus of the ides immanent in nervous activity, Bulletin of Mathematical Biophysics, 5: 115–133.

Mizraji, E. (1989) Context-dependent associations in linear distributed memories, Bulletin of Mathematical Biology, 50: 195–205.

Mizraji, E. (1994) Lógicas vectoriales: una aproximación a las bases neurales del pensamiento lógico, Galileo (Uruguay), Segunda época, 10: 3–16.

Mizraji, E. (1999) El Segundo Secreto de la Vida Trilce, Montevideo.

Mizraji, E. (2008) Neural memories and search engines, International Journal of General Systems, 37: 715–732.

Monod, J. (1972) Chance and Necessity, Vintage Books, New York.

Pomi, A., Mizraji, E. (1999) Memories in contexts, BioSystems, 50: 173–188.

Rosenblatt, F. (1958) The Perceptron: A probabilistic model for information storage and organization in the Brain, Psychological Reviews, 65: 386–408.

Valle-Lisboa, J.C., Pomi, A., Mizraji, E. (2023) Multiplicative processing in the modeling of cognitive activities in large neural networks, Biophysical Reviews, 15: 767–785.

Part 2: **Vector logic**

Chapter 4
The matrix operators of logic

The case described in the final part of the previous chapter shows that the translation of the truth tables shown in Chapter 1 into the language of matrix operators is immediate (Mizraji 1992, 1996). To introduce matrix representation, we start by defining veritative values as vectors. Let us consider two basic vectors, s and n. The first, s, is a column vector that corresponds to the concept "true" and the second, n, a column vector that represents the concept "false." We will assume that both vectors are of dimension Q. In this formalism, the set τ of truth values is

$$\tau = \{s, \ n\}.$$

We will assume, in the first instance, that the vectors s and n are orthonormal. Monadic and dyadic operators, as in classical scalar logic, are generated by the following applications:

$$\mu : \tau \longrightarrow \tau,$$

$$\delta : \tau \times \tau \longrightarrow \tau.$$

In this formalism, the structure of monadic operators is given by the equation

$$\mu(a, b) = as^T + bn^T; \quad a, b \in \tau. \tag{4.1}$$

These operators are square matrices of order $Q \times Q$ and with the formal structure of the associative memories described in eq. (3.5).

The general equation for dyadic operators is

$$\delta(a, b, c, d) = a(s \otimes s)^T + b(s \otimes n)^T + c(n \otimes s)^T + d(n \otimes n)^T; \quad a, b, c, d \in \tau. \tag{4.2}$$

These dyadic operators are rectangular matrices of order $Q \times Q^2$ and have the structure of associative memories with multiplicative contexts.

4.1 Monadic operators

a) *Identity*

$$I = ss^T + nn^T.$$

Note that by the orthonormality of the pair s, n, it is $Is = s\langle s, s\rangle + n\langle n, s\rangle = s$ and $In = n$. (It should be noted that this identity operator is not generally the same as the identity matrix of linear algebra, although it can be for certain pairs s, n.)

https://doi.org/10.1515/9783112230053-005

b) *Negation*

$$N = ns^T + sn^T.$$

Hence, $Ns = n$ and $Nn = s$.

Let us also note, as it will be useful to us in further calculations, that $s^T N = n^T$ and that $n^T N = s^T$.

c) *Affirmation*

$$K = ss^T + sn^T.$$

This operator only states: $Ks = Kn = s$.

d) *Refusal*

$$M = ns^T + nn^T.$$

This operator only denies: $Ms = Mn = n$.

The fact that the four monadic operations can be represented as matrices allows us to see clearly that we are dealing with a theory of operators, where they are susceptible to being operated on each other, in the absence of variables. Thus, we can see, through the operations of linear algebra, that the double negative generates an identity. Let us confirm this with a detailed calculation assuming the orthonormality of $\{s, n\}$:

$$NN = N^2 = \left(ns^T + sn^T\right)\left(ns^T + sn^T\right) =$$
$$\left(ns^T ns^T + ns^T sn^T + sn^T ns^T + sn^T sn^T\right) =$$
$$n\langle s, n\rangle s^T + n\langle s, s\rangle n^T + s\langle n, n\rangle s^T + s\langle n, s\rangle n^T =$$
$$nn^T + ss^T = I.$$

Similarly, identity can be shown to be idempotent: $I^2 = I$.

It can also be proved that pre- and post-multiplication by I leaves the other monadic operators unaltered. The following expressions show links between monadic operators:

a) $NN = I;\ NK = M;\ NM = K$;
b) $KN = KK = KM = K$;
c) $MN = MK = MM = M$.

Let us conclude by exemplifying the numerical structure of I and N for two pairs of two-dimensional vectors corresponding to two different orthonormal sets $\{s, n\}$.

Example 1

$$s = [1 \quad 0]^T; \quad n = [0 \quad 1]^T;$$

$$I = \begin{bmatrix} 1 & 0 \\ 0 & 1 \end{bmatrix}; \quad N = \begin{bmatrix} 0 & 1 \\ 1 & 0 \end{bmatrix}.$$

Example 2

$$s = \left(1/\sqrt{2}\right)[1 \quad 1]^T; \quad \left(1/\sqrt{2}\right)[1 \quad -1]^T;$$

$$I = \begin{bmatrix} 1 & 0 \\ 0 & 1 \end{bmatrix}; \quad N = \begin{bmatrix} 1 & 0 \\ 0 & -1 \end{bmatrix}.$$

Note that for these examples the identity operator coincides with the identity matrix, which is not a general fact. On the contrary, negations are different, although both are involutive and satisfy $N^2 = I$.

4.2 Dyadic operators

We now show how matrices of order $Q \times Q^2$ correspond to the dyadic logical tables in Chapter 1.

a) *Conjunction*

$$C = s \ (s \otimes s)^T + n \ (s \otimes n)^T + n \ (n \otimes s)^T + n \ (n \otimes n)^T.$$

Calculating, it turns out that

$$C(s \otimes s) = s,$$
$$C(s \otimes n) = C(n \otimes s) = C(n \otimes n) = n.$$

b) *Disjunction*

$$D = s \ (s \otimes s)^T + s \ (s \otimes n)^T + s \ (n \otimes s)^T + n \ (n \otimes n)^T$$
$$D(s \otimes s) = D(s \otimes n) = D(n \otimes s) = s,$$
$$D(n \otimes n) = n.$$

c) *Implication*

$$L = s \ (s \otimes s)^T + n \ (s \otimes n)^T + s \ (n \otimes s)^T + s \ (n \otimes n)^T,$$

$$L(s\otimes s) = L(n\otimes s) = L(n\otimes n) = s,$$
$$L(s\otimes n) = n.$$

d) *Equivalence*

$$E = s\ (s\otimes s)^T + n\ (s\otimes n)^T + n\ (n\otimes s)^T + s\ (n\otimes n)^T,$$

$$E(s\otimes s) = E(n\otimes n) = s,$$
$$E(s\otimes n) = E(n\otimes s) = n.$$

e) *Exclusive-or*

$$X = n\ (s\otimes s)^T + s\ (s\otimes n)^T + s\ (n\otimes s)^T + n\ (n\otimes n)^T,$$

$$X(s\otimes n) = X(n\otimes s) = s,$$
$$X(s\otimes s) = X(n\otimes n) = n.$$

f) *Sheffer's connective*

$$S = n\ (s\otimes s)^T + s\ (s\otimes n)^T + s\ (n\otimes s)^T + s\ (n\otimes n)^T,$$

$$S(s\otimes s) = n,$$
$$S(s\otimes n) = S(n\otimes s) = S(n\otimes n) = s.$$

g) *Peirce's connective*

$$P = n\ (s\otimes s)^T + n\ (s\otimes n)^T + n\ (n\otimes s)^T + s\ (n\otimes n)^T,$$

$$P(s\otimes s) = P(s\otimes n) = P(n\otimes s) = n,$$
$$P(n\otimes n) = s.$$

Let us exemplify the numerical structure of L and X for the two bases used earlier.

Example 1

$$s = [1 \quad 0]^T;\ n = [0 \quad 1]^T;$$

$$L = \begin{bmatrix} 1 & 0 & 1 & 1 \\ 0 & 1 & 0 & 0 \end{bmatrix};\quad X = \begin{bmatrix} 0 & 1 & 1 & 0 \\ 1 & 0 & 0 & 1 \end{bmatrix}.$$

Example 2

$$s = \left(1/\sqrt{2}\right)[1 \quad 1]^T;\ \left(1/\sqrt{2}\right)[1 \quad -1]^T;$$

$$L = \frac{1}{\sqrt{2}} \begin{bmatrix} 2 & 0 & 0 & 0 \\ 1 & 1 & -1 & 1 \end{bmatrix};\quad X = \frac{1}{\sqrt{2}} \begin{bmatrix} 2 & 0 & 0 & 0 \\ 0 & 0 & 0 & -2 \end{bmatrix}.$$

Note that, except in the case of implication, the truth vectors commute; thus, for $u, v \in \tau$ we have

$$C(u \otimes v) = C(v \otimes u),$$
$$D(u \otimes v) = D(v \otimes u), \text{ etc.}$$

The negation N generates the following equalities:

$$X = NE,$$
$$S = NC = D(N \otimes N),$$
$$P = ND = C(N \otimes N).$$

The logical equivalence between implication and disjunction is well known:

$$p \Rightarrow q \equiv \neg p \vee q.$$

The matrix version of this equivalence is an equality between vectors, given by

$$L(u \otimes v) = D(N u \otimes v) ; \quad u, v \in \tau.$$

But note that from the properties of Kronecker's product we can, by factoring, express this equation as follows:

$$L(u \otimes v) = D(N \otimes I)(u \otimes v).$$

We ask: are matrices L and $D(N \otimes I)$ equal? The answer is yes, and it is tested by calculating the second one from the definitions and finding that it generates the matrix that defines L. For example, for the first term of the definition of D, multiplied by $N \otimes I$, it results:

$$s \, (s \otimes s)^T (N \otimes I) = s \, (s^T N \otimes s^T I) = s \, (n^T \otimes s^T) = s \, (n \otimes s)^T,$$

and thus, the different terms that define L are reconstructed. Consequently, we have the following equality between operators:

$$L = D(N \otimes I).$$

This important result shows a crucial property of this approach: logical equivalences, in many situations, correspond to equalities between operators, *without the vectors on which they operate being involved in these equalities*. We will see in a later chapter how fundamental this property is.

With a further step of complexity, but also through direct calculations, it can be shown that De Morgan's relations produce matrix identities. Putting negation before the standard versions we present in Chapter 1 (in the item "equivalential tautologies") result two other well-known expressions:

De Morgan's law 1: $p \wedge q \equiv \neg(\neg p \vee \neg q)$;
De Morgan's law 2: $p \vee q \equiv \neg(\neg p \wedge \neg q)$.

If we analyze this with the format of operators, factor the variables, and perform direct calculations from the definitions, we find the following matrix equations:

De Morgan's law 1': $C = ND(N \otimes N)$;
De Morgan's law 2': $D = NC(N \otimes N)$.

Therefore, in this format, De Morgan's laws are also equalities between operators, where vector variables disappear. On the other hand, as a mere exercise in consistency, we show the passage from the first version 1' to the second 2'. This is done with direct algebraic calculations. Thus, pre-multiplying C by N and post-multiplying by $N \otimes N$, and then, expressing C using De Morgan's Law 1', we have

$$NC(N \otimes N) = NND(N \otimes N)(N \otimes N) =$$
$$ID(N^2 \otimes N^2) = ID(I \otimes I) = D.$$

This tells us that if we had not determined the matrix equality corresponding to De Morgan's law 2', we could have calculated it from the first within the rules of matrix algebra.

Now we are going to see an important tautology that does not give rise to equality between operators, but that in this formalism is elegantly demonstrated by appealing to the operations between matrices and vectors. It is the law of contraposition, which in classical logic has this presentation:

$$(p \Rightarrow q) \equiv (\neg q \Rightarrow \neg p).$$

For a couple of vectors $u, v \in \tau$, we write

$$L(u \otimes v) = D(N \otimes I)(u \otimes v) = D(Nu \otimes Iv).$$

Taking into account that $I = NN$ and the commutativity of the disjunction, from the last expression the following chain of equalities arises:

$$D(Nu \otimes Iv) = D(Nu \otimes NNv) = D(NNv \otimes Nu) =$$
$$D(NNv \otimes INu) = D(N \otimes I)(Nv \otimes Nu) = L(Nv \otimes Nu).$$

Consequently, $L(u \otimes v) = L(Nv \otimes Nu)$.

Finally, we have suggested that class logic and propositional logic in this formalism of operators can converge on the same matrices (Mizraji 1992; Mizraji and Lin 2011). This is argued from a vector definition of the characteristic functions of sets. Let us assume that it is possible to associate each of the elements of a set of objects with a concept encoded by a vector. Let us define a vector function $f(z, S)$ which defines whether the object to which the vector z is associated belongs to the set of objects S. This generates the following decision rule:

$$f(z,S) = \begin{cases} s & \text{if } z \in S \\ n & \text{if } z \notin S \end{cases},$$

where $s, n \in \tau$. Note that this function is equivalent to a logical identity I: if z belongs to S, the output is s; if z does not belong to S, the output is n. So, if S is the complement of a set A, or $S = A \cap B$, or $S = A \cup B$, we have a suggestive result, which links the vector version of a characteristic function of the complement, intersection, and union sets with the matrix operators N, C, and D:

$$f(z, \bar{A}) = N,$$

$$f(z, A \cap B) = C,$$

$$f(z, A \cup B) = D.$$

4.3 A note on the exclusive-or

The exploration of ways to compute the exclusive-or, XOR, has promoted much research in the neural domain, and from the 1943 model of McCulloch and Pitts to the development of models with hidden layers, XOR became an important logical object for researchers in neural network models. As we pointed out at the beginning, the matrix formalism that we present in this book was also inspired by the ease of computation of XOR by matrix associative memories. But what is most curious about XOR is that, as defined by its truth table or the X-matrix, it cannot be a model of the exclusive disjunction that our cognition uses. The problem arises from the logical decision that humans make in the face of the following type of eventuality. Suppose we are faced with this proposition: "Yesterday, at exactly 8:06 a.m., I was at home having breakfast or on the road returning to Montevideo, or in a hotel in Buenos Aires." The only triples with possible truth value are (100), (010) and (001). Other triples containing more than only 1 cannot bet true in the cognitive domain because they would give true to the fact of being at the same time in two or three different places. Now, if we evaluate the expressions using the XOR truth table XOR[XOR(1,1),1] and XOR[1,XOR(1,1)], it turns out that they are both true. And in its matrix version we have

$$X[X(s \otimes s) \otimes s] = X[s \otimes X(s \otimes s)] = s.$$

The conclusion that is imposed is that this exclusive-or does not compute the same as the exclusive-or of human cognition. The concatenated XOR is an indicator of the parity of a binary sequence, giving 1 if there is an odd number of 1's and 0 if the number of 1's is even.

We have presented an alternative for the construction of an exclusive disjunction compatible with cognitive computations (Mizraji, Pomi, Reali, and Valle Lisboa 2003;

Mizraji and Lin 2011). This alternative (inspired by the existence of recursive neural modules that act in parallel), consists of creating an evaluation system that computes for $t+1$ and in parallel the alternatives presented at time t, and arrives at the final result by a simple disjunction of the set of results. It is a dynamic XOR because we create a dynamic system for its calculation. We illustrate this using the matrix format in the evaluation of dynamic XOR (X_{din}) for two and three variables.

Dynamic XOR: two variables:

$$\begin{cases} u(t+1) = C[u(t) \otimes Nv(t)] \\ v(t+1) = C[v(t) \otimes Nu(t)] \end{cases}$$

$$X_{\text{din}}(u,v) = D[u(t+1) \otimes v(t+1)].$$

Dynamic XOR: three variables:

$$\begin{cases} u(t+1) = C[C(u(t) \otimes Nv(t)) \otimes Nw(y)] \\ v(t+1) = C[C(v(t) \otimes Nw(t)) \otimes Nu(y)] \, , \\ w(t+1) = C[C(w(t) \otimes Nu(t)) \otimes Nv(y)] \end{cases}$$

$$X_{\text{din}}(u,v,w) = D[D(u(t+1) \otimes v(t+1)) \otimes w(t+1)].$$

This seems like an exaggeratedly complicated way of executing the XOR, but it has the virtue that for 2 variables it reconstructs the dyadic XOR, but for more than two variables it meets what we would demand of a cognitive XOR. The tables are described in the references, and the reader can construct them without difficulty. What here are equations expressed in parallel, a system with memory that retains evaluations in $t+1$, can produce it by recursions based on conjunction and disjunction.

4.4 Normal vectors but not necessarily orthogonal

Previously, we have expressed the logical operators assuming that the pair s, n was orthonormal. Now, we are going to retain the normality of the vectors while assuming that they can be non-orthogonal. Therefore, the pair s, n is formed by pairs of column vectors of modulus 1, whose scalar products are given by the following equations (Mizraji 1996):

$$\langle s, s \rangle = \langle n, n \rangle = 1,$$
$$\langle s, n \rangle = \langle n, s \rangle = \varepsilon,$$

with $\varepsilon \neq |1|$, which means that since they cannot be parallel, these vectors s,n are linearly independent. Now our τ truth value set contains this new vector pair.

Here, we will follow the strategy of Kohonen (1977) in his derivation of the structure of matrix memories and we will apply it to derive monadic and dyadic operators.

4.4.1 Monadic operators

Let us call any of the four monadic operators U and propose the search for U from the following matrix equation:

$$[a \ b] = U[s \ n] \; ; \quad a, b \in \tau. \tag{4.3}$$

So, the solution can be expressed in terms of the Moore–Penrose pseudoinverse:

$$U = [a \ b][s \ n]^+. \tag{4.4}$$

The equality sign is due to the fact that the linear independence of s and n ensures that an exact solution can be calculated using the formula

$$A^+ = (A^T A)^{-1} A^T.$$

Performing the calculations, the following equation is obtained:

$$[s \ n]^+ = \frac{1}{1-\varepsilon^2} \begin{bmatrix} 1 & -\varepsilon \\ -\varepsilon & 1 \end{bmatrix} \begin{bmatrix} s^T \\ n^T \end{bmatrix}. \tag{4.5}$$

Defining

$$y = \frac{1}{1-\varepsilon^2}(s - \varepsilon n) \; ; \quad z = \frac{1}{1-\varepsilon^2}(n - \varepsilon s) \tag{4.6}$$

we get

$$[s \ n]^+ = [y \ z]^T. \tag{4.7}$$

In this way, the four monadic operators can express themselves as

$$U = ay^T + bz^T. \tag{4.8}$$

Let us note that $\langle y, s \rangle = \langle z, n \rangle = 1$ and that $\langle y, n \rangle = \langle z, s \rangle = 0$.

Defining the matrix $V = [y \ z]$ we can represent monadic operators as

$$U = [a \ b] V^T. $$

4.4.2 Dyadic operators

In this case, the problem statement has this structure

$$[a \ b \ c \ d] = T([s \ n] \otimes [s \ n]); \quad a, b, s, d \in \tau, \tag{4.9}$$

where T is any of the 16 dyadic operators.

We will define Kronecker's product for partitioned matrices $[g \quad h]$ as follows.

$$[g \quad h] \otimes [g \quad h] = [g \otimes [g \quad h] \quad h \otimes [g \quad h]] =$$
$$[(g \otimes g) \quad (g \otimes h) \quad (h \otimes g) \quad (h \otimes h)].$$

On this basis, the general solution using the pseudoinverse is

$$T = [a \quad b \quad c \quad d]([s \quad n] \otimes [s \quad n])^+. \tag{4.10}$$

Using the following link between pseudoinverse and Kronecker's product:

$$(A \otimes B)^+ = A^+ \otimes B^+.$$

Equation (4.10) can be written like this:

$$T = [a \quad b \quad c \quad d]([s \quad n]^+ \otimes [s \quad n]^+). \tag{4.11}$$

From the previous results and definitions, it is clear that

$$T = [a \quad b \quad c \quad d]([y \quad z] \otimes [y \quad z])^T. \tag{4.12}$$

Developing this expression, we find a general representation for dyadic operators:

$$T = a \ (y \otimes y)^T + b \ (y \otimes z)^T + c \ (z \otimes y)^T + d \ (z \otimes z)^T. \tag{4.13}$$

Defining an F matrix as

$$F = [(y \otimes y) \ (y \otimes z) \ (z \otimes y) \ (z \otimes z)]$$

results in

$$T = [a \quad b \quad c \quad d] \ F^T. \tag{4.14}$$

This expression is interesting because it gives us an alternative representation also for classical dyadic operators. If the vectors s, n are orthonormal, then $\varepsilon = 0$. In that case we define $H = F$ for $\varepsilon = 0$ as follows:

$$H = [(s \otimes s) \ (s \otimes n) \ (n \otimes s) \ (n \otimes n)].$$

In this case, it is

$$T = [a \quad b \quad c \quad d] \ H^T. \tag{4.15}$$

In this way, the matrix operators conjunction C and disjunction D can be represented as follows:

$$C = [s \ n \ n \ n] \ H^T,$$
$$D = [s \ s \ s \ n] \ H^T.$$

We can assimilate this representation to the one used by Wittgenstein in the logical tables of item 5.101 of the *Tractatus* (see Figure 1.1).

4.5 Kronecker polynomials

The structures that we are going to define as "Kronecker polynomials" were created to demonstrate that triadic logical operators (in the sense of being dependent on three variables) or of greater complexity, could be represented by a polynomial expansion whose coefficients are the monadic matrices and the multiplications involved are performed by Kronecker products (Mizraji 1996). This formalism shows that the monadic matrices I, N, K, and M can be metaphorically thought of as "atoms" from whose combination the most complex operators arise.

4.5.1 First-degree Kronecker polynomials

The construction of these polynomials is based on the following property of the Kronecker product operating on column vectors:

$$u \ (v \otimes w)^T = uv^T \otimes w^T. \tag{4.16}$$

Proof

$$(u \otimes 1)(v^T \otimes w^T) = uv^T \otimes 1.w^T = uv^T \otimes w^T.$$

Using this property, a dyadic operator

$$T = a \ (s \otimes s)^T + b \ (s \otimes n)^T + c \ (n \otimes s)^T + d \ (n \otimes n)^T; \quad a, b, c, d \in \tau$$

can be expressed as

$$T = U_1 \otimes s^T + U_2 \otimes n^T \tag{4.17}$$

with $U_1 = as^T + cn^T$ and $U_2 = bs^T + dn^T$. We call a structure like (4.17) the first-degree Kronecker polynomial. As you can see, $U_1, U_2 \in \{I, N, K, M\}$.

These polynomials generate the following representations for the dyadic operators we defined earlier:

$$C = I \otimes s^T + M \otimes n^T,$$

$$D = K \otimes s^T + I \otimes n^T,$$

$$L = K \otimes s^T + N \otimes n^T,$$

$$E = I \otimes s^T + N \otimes n^T,$$

$$X = N \otimes s^T + I \otimes n^T,$$

$$S = N \otimes s^T + K \otimes n^T,$$

$$P = M \otimes s^T + N \otimes n^T.$$

In this way, the structure of these dyadic operations is completely defined by their monadic coefficients. We can represent it in abbreviated form in this way: $[U_1, U_2)$.

Similarly, and based on a property of the Kronecker product analogous to that shown in eq. (4.16), it is possible to define left-hand first-degree Kronecker polynomials, represented by the following general expression:

$$T = s^T \otimes V_1 + n^T \otimes V_2 \equiv (V_1, V_2]$$

$(V_1, V_2 \in \{I, N, K, M\})$. It can be verified that for the operators C, D, E and their negations S, P, X the left and right polynomials coincide, and $[U_1, U_2) = (U_1, U_2]$ On the other hand, for the implication, we have

$$L = [KN) = (IK].$$

4.5.2 Second-degree Kronecker polynomials

If we define a logical function of three variables, the associated matrix would have this structure:

$$G = h_1(s \otimes s \otimes s)^T + h_2(s \otimes s \otimes n)^T + h_3(s \otimes n \otimes s)^T + h_4(s \otimes n \otimes n)^T +$$
$$h_5(n \otimes s \otimes s)^T + h_6(n \otimes s \otimes n)^T + h_7(n \otimes n \otimes s)^T + h_8(n \otimes n \otimes n)^T; \quad h_i \in \tau. \tag{4.18}$$

In a similar way to what is established for dyadic functions, in this case, we can represent these matrices by polynomials

$$G = U_1 \otimes (s \otimes s)^T + U_2 \otimes (s \otimes n)^T + U_3 \otimes (n \otimes s)^T + U_4 \otimes (n \otimes n)^T, \tag{4.19}$$

a structure that we call second-degree Kronecker polynomials. This expression comes from applying a variant of the property (4.16) noting that of the eight terms of (4.18) there are two whose last Kronecker factor is $(s \otimes s)^T$, two with $(s \otimes n)^T$, etc. Consequently, the values of the monadic coefficients are defined by the following equations:

$$U_1 = h_1 s^T + h_5 n^T; \; U_2 = h_2 s^T + h_6 n^T;$$

$$U_3 = h_3 s^T + h_7 n^T; \; U_4 = h_4 s^T + h_8 n^T.$$

We will compactly symbolize these second-degree polynomials by the expression $G = [U_1 \; U_2 \; U_3 \; U_4)$.

In a similar way, the left and middle second-degree Kronecker polynomials are defined:

$$(V_1 V_2 V_3 V_4] = (s \otimes s)^T \otimes V_1 + (s \otimes n)^T \otimes V_2 + (n \otimes s)^T \otimes V_3 + (n \otimes n)^T \otimes V_4,$$

$$(W_1 W_2 W_3 W_4) = s^T \otimes W_1 \otimes s^T + s^T \otimes W_2 \otimes n^T + n^T \otimes W_3 \otimes s^T + n^T \otimes W_4 \otimes n^T.$$

On this basis, we will now show the situations that generate these second-degree polynomials. Let us consider the following logical identities concerning the laws of associativity between dyadic connectives:

$$(a1) \quad p \wedge (q \wedge r) = (p \wedge q) \wedge r,$$

$$(a2) \quad p \vee (q \vee r) = (p \vee q) \vee r,$$

$$(a3) \quad [p \equiv (q \equiv r)] = [(p \equiv q) \equiv r].$$

If we transfer these equivalences to the matrix domain, for $u, v, w \in \tau$, it is

$$(b1) \quad C \, [u \otimes C(v \otimes w)] = C \, [\, C(u \otimes v) \otimes w],$$

$$(b2) \quad D \, [u \otimes D(v \otimes w)] = D \, [\, D(u \otimes v) \otimes w],$$

$$(b3) \quad E \, [u \otimes E(v \otimes w)] = E \, [\, E(u \otimes v) \otimes w].$$

Let us also consider the following identity, associated with the tautology called "import law"

$$(a4) \quad [p \Rightarrow (q \Rightarrow r)] = [(p \wedge q) \Rightarrow r].$$

Its matrix representation is

$$(b4) \quad L[u \otimes L(v \otimes w)] = L[C(u \otimes v) \otimes w].$$

Thanks to Kronecker's product properties, it is possible to separate variables and operators. We illustrate this for case (b1):

$$C(I \otimes C)(u \otimes v \otimes w) = C(C \otimes I)(u \otimes v \otimes w).$$

This suggests the following equalities between operators:

(c1) $C(I \otimes C) = C(C \otimes I)$,

(c2) $D(I \otimes D) = D(D \otimes I)$,

(c3) $E(I \otimes E) = E(E \otimes I)$,

(c4) $L(I \otimes L) = L(C \otimes I)$,

which are in fact demonstrable directly from the definition of each of the dyadic operators involved, using extensive calculations. A more compact way to prove these equalities is by using the representation of dyadic operators as Kronecker polynomials (4.17) and the format described in eq. (4.15). For these calculations, the following two theorems (demonstrated in Mizraji (1996)) are useful:

Theorem 4.1. Let the operators be dyadic $A = A_1 \otimes s^T + A_2 \otimes n^T$ and $B = [b_1\ b_2\ b_3\ b_4]$ H^T. Then, $A(I \otimes B) = [U_1\ U_2\ U_3\ U_4]$ with $U_i = A_1$, if $b_i = s$ and $U_i = A_2$ if $b_i = n$ (i = 1, 2, 3, 4).

Theorem 4.2. If $A = A_1 \otimes s^T + A_2 \otimes n^T$ and $B = B_1 \otimes s^T + B_2 \otimes n^T$, then $A(B \otimes I) = [U'_1\ U'_2\ U'_3\ U'_4)$ with $U'_1 = A_1 B_1$, $U'_2 = A_2 B_1$, and $U'_3 = A_1 B_2$ $U'_4 = A_2 B_2$.

The equalities c1–c4 generate identical second-degree Kronecker polynomials, as shown from these theorems, which is a way of proving the equalities. We now show the shape of the polynomials that are associated with these equalities:

(i) $[I\ M\ M\ M) \ = \ C(I \otimes C) \ = \ C(C \otimes I)$,

(ii) $[K\ K\ K\ I) \ = \ D(I \otimes D) \ = \ D(D \otimes I)$,

(iii) $[I\ N\ N\ I) \ = \ E(I \otimes E) \ = \ E(E \otimes I)$.

Note that the polynomial (iii) also corresponds to an analogous equality, where instead of E the operator is X:

(iii)′ $[I\ N\ N\ I) \ = \ X(I \otimes X) \ = \ X(X \otimes I)$.

The matrix version of the export law has this polynomial expression:

(iv) $[K\ N\ K\ K) \ = \ L(I \otimes L) \ = \ L(C \otimes I)$.

Let us now look at an analysis of the most classical form of the hypothetical syllogism

$$[(p \Rightarrow q) \wedge (q \Rightarrow r)] \Rightarrow (p \Rightarrow r).$$

If we transfer this expression into the formalism of matrix operators, it results:

$$h_{SH} = L\{C\,[\,L(u\otimes v)\otimes L(v\otimes w)]\otimes L(u\otimes w)\}.$$

If we separate operators from variables, it turns out that

$$h_{SH} = L[C(L\otimes L)\otimes L]\,(u\otimes v\otimes v\otimes w\otimes u\otimes w).$$

But let us look at the following chain of equalities:

$$C(L\otimes L)\otimes L{=}C(L\otimes L)\otimes I\ L{=}(C\otimes I)L^{[3]},$$

where $L^{[3]}$ represents Kronecker's triple power $L\otimes L\otimes L$. Consequently, the matrix of the hypothetical syllogism shown can be written $L(C\otimes I)L^{[3]}$. But if we consider equality (c4), we can write the following equality:

$$L(C\otimes I)L^{[3]} = L(I\otimes L)L^{[3]}.$$

Reconstructing the meaning of the second member of this equation in terms of classical logic, this formula results:

$$(p \Rightarrow q) \Rightarrow [(q \Rightarrow r) \Rightarrow (p \Rightarrow r)],$$

which is another form of the hypothetical syllogism that can be found in item 6.32 of the of Bochenski's text (1959, p. 23). Bochensky, along with the classical version, includes this syllogism in Polish notation. If we translate the original Polish symbols and use the letters of our matrix notation in italics, their Polish version of this syllogism is:

$$LLpqLLqrLpr.$$

If we now express the new version obtained with the formalism of operators, it results:

$$L\{L(u\otimes v)\otimes L[L(v\otimes w)\otimes L(u\otimes w)]\}.$$

Note that if we dispense with parentheses and signs, the Polish version shows the same succession of logical connectives and variables as the matrix version.

4.5.3 Links between the monadic coefficients of Kronecker polynomials

We have just seen that complex logical expressions, such as the hypothetical syllogism we have been discussing, can take different forms based on dyadic functions, but all of them represent the same logical operation (a case of mathematical synonymy). In these cases, Kronecker polynomials and their monadic coefficients constitute invariants of that logical function through their various "synonyms.". Here we show the relationships between these monadic matrices (Mizraji 1996).

a) Equivalences between U_i and V_i.
The coefficients of the right and left first-degree Kronecker polynomials are linked by the following equations:

$$V_1 = U_1 ss^T + U_2 sn^T,$$

$$V_2 = U_1 ns^T + U_2 nn^T.$$

In the case of implication, we had $L = [K, N) = (I, K]$. The previous formula allows the following calculation:

$$V_1 = Kss^T + Nsn^T = ss^T + nn^T = I,$$

$$V_2 = Kns^T + Nnn^T = ss^T + sn^T = K.$$

b) Equivalences between U_i, V_i, and W_i.
The coefficients associated with the left and middle second-degree Kronecker polynomials can be directly calculated from the monadic coefficients of the right polynomial using the following equations:

$$
\begin{aligned}
V_1 &= U_1 ss^T + U_2 sn^T, & W_1 &= U_1 ss^T + U_3 sn^T, \\
V_2 &= U_3 ss^T + U_4 sn^T, & W_2 &= U_2 ss^T + U_4 sn^T, \\
V_3 &= U_1 ns^T + U_2 nn^T, & W_3 &= U_1 ns^T + U_3 nn^T, \\
V_4 &= U_3 ns^T + U_4 nn^T, & W_4 &= U_2 ns^T + U_4 nn^T.
\end{aligned}
$$

We will see in a later chapter that these second-degree Kronecker polynomials unexpectedly turned out to be an instrument through which to explore the dynamical properties of the elementary cellular automata created by Stephen Wolfram.

4.6 The square root of negation as a powerful computational tool

In the field of quantum computing research, the square root of negation emerged as an important operator (Hayes 1995; Deutsch et al. 2000). In this context, the negation operator N is a square matrix of order 2, supported by the two-dimensional vectors that are the qubits, the physical basis of quantum computing models. The work of Deutsch et al. (2000) shows how the square root of N (which is an operator in the complex domain) expands computational power so that analyzing the output of a single input gathers global information, inaccessible in the non-quantum domain. Later, this operator was generalized for Q-dimensional vectors (Mizraji 2008) and we show its structure below.

Given the formulas we have seen for identity and negation,

$$I = nn^T + ss^T \text{ and } N = sn^T + ns^T,$$

there are two square roots of N given by the following equations:

$$A = \left(\sqrt{N}\right)_1 = \frac{1}{2}(1+i)\,I + \frac{1}{2}(1-i)\,N, \qquad (4.20)$$

$$B = \left(\sqrt{N}\right)_2 = \frac{1}{2}(1-i)\,I + \frac{1}{2}(1+i)\,N, \qquad (4.21)$$

with $= \sqrt{-1}$.

These roots have interesting properties, which we demonstrate below:

$$A^2 = N;\ B^2 = N;\ AB = BA = I.$$

There is a clear analogy with the two square roots of −1. For instance, the positive root $+\left(\sqrt{-1}\right)$ has a matrix version

$$\left(\sqrt{N}\right)_1 = IA,$$

and $-\left(\sqrt{-1}\right)$ has its matrix version in

$$\left(\sqrt{N}\right)_2 = NA.$$

It is immediate that $NA = B$.

To show an elementary numerical example, we will use the two sets of vectors used previously:

Set 1: $s = \begin{bmatrix} 1 \\ 0 \end{bmatrix}$, $n = \begin{bmatrix} 0 \\ 1 \end{bmatrix}$; Set 2: $s = \frac{1}{\sqrt{2}}\begin{bmatrix} 1 \\ 1 \end{bmatrix}$, $n = \frac{1}{\sqrt{2}}\begin{bmatrix} 1 \\ -1 \end{bmatrix}$.

The values of $\sqrt{N}$ for Set 1 and Set 2 are:

Set 1: $I = \begin{bmatrix} 1 & 0 \\ 0 & 1 \end{bmatrix}$, $N = \begin{bmatrix} 0 & 1 \\ 1 & 0 \end{bmatrix}$, $A = \begin{bmatrix} \frac{1}{2}(1+i) & \frac{1}{2}(1-i) \\ \frac{1}{2}(1-i) & \frac{1}{2}(1+i) \end{bmatrix}$, $B = \begin{bmatrix} \frac{1}{2}(1-i) & \frac{1}{2}(1+i) \\ \frac{1}{2}(1+i) & \frac{1}{2}(1-i) \end{bmatrix}$,

Set 2: $I = \begin{bmatrix} 1 & 0 \\ 0 & 1 \end{bmatrix}$, $N = \begin{bmatrix} 1 & 0 \\ 0 & -1 \end{bmatrix}$, $A = \begin{bmatrix} 1 & 0 \\ 0 & i \end{bmatrix}$, $B = \begin{bmatrix} 1 & 0 \\ 0 & -i \end{bmatrix}$.

An area where the square root of N operator allowed for an interesting result is in the investigation of counterfactual implications (Mizraji 2021a). It is found that for a counterfactual proposition, where the propositions and their truth values refer to a virtual past (e.g., "if Roger had been on defense then we would not have lost the game"), the square roots of N produce outputs weighted by imaginary coefficients. So if

$$L^c(p^* \otimes q^*)$$

is a counterfactual implication, where p^* and q^* are the positive evaluations assigned to the propositions of that virtual situation, then explicit virtualization occurs if one proceeds as follows:

$$AL^c(s \otimes s) = As \ and \ As = \frac{1}{2}(1+i) \ s + \frac{1}{2}(1-i) \ n.$$

Subsequently, the two virtual options will have to be subjected to empirical scrutiny to assess whether the counterfactual idea was good and whether plans need to be changed (in our example, putting Roger on defense in the next match).

In Mizraji (2021b), based on the calculation by Deutsch et al. (2000), it is shown how a single input of a truth table, preceded by the square root of N operator, displays the remaining values of the table and allows them to be uniquely identified by projecting them onto the real and complex hyperplanes.

References

Bochenski, J.M. (1959) A precis of Mathematical Logic, Springer, Dordrecht.

Deutsch, D., Ekert, A., Lupacchini, R. (2000) Machines, logic and quantum physics. The Bulletin of Symbolic Logic, 6: 265–283.

Hayes, B. (1995) The square root of NOT. American Scientist, 83: 304–308.

Kohonen, T. (1977) Associative Memory: A System-Theoretical Approach, Springer, New York.

Mizraji, E. (1992) Vector logic: The matrix-vector representation of logical calculus. Fuzzy Sets and Systems, 50: 179–185.

Mizraji, E. (1996) The operators of vector logic, Math. Log. Quart, 42: 27–40.

Mizraji, E., Pomi, A., Reali, F., Valle-Lisboa, J.C. (2003) Disyunciones dinámicas, in Procesos biofísicos complejos, pp. 29–48 (J.A. Hernández, A. Pomi, Eds), pp. 29–48 DIRAC, Montevideo.

Mizraji, E. (2008) Vector logic: A natural algebraic representation of the fundamental logical gates. Journal of Logic and Computation, 18: 97–121.

Mizraji, E., Lin, J. (2011) Logic in a dynamic brain, Bull. Math. Biol. 73: 373–397.

Mizraji, E. (2021a) Vector logic allows counterfactual virtualization by the square root of NOT. Logic Journal of the IGPL, 29: 859–870.

Mizraji, E. (2021b) Generalized "Square roots of Not" matrices, their application to the unveiling of hidden logical operators and to the definition of fully matrix circular Euler functions. Preprint in arXiv:2107.06067v5 (https://arxiv.org/abs/2107.06067)

Chapter 5
Binary matrix operators and uncertain logical values

In 1920, in Poland, Jan Lukasiewicz published a two-page article in a local philosophy journal whose title, translated into English, is "On Trivalent Logic." In that short article, Lukasiewicz added a third truth value, placing it somewhere between truth and falsehood. In that article, he assigned this third value to "propositions that are neither true nor false" and in his nomenclature of 1 and 0 for truth or falsehood, respectively, he assigns the symbol 1/2. In this first work, where he still does not use the Polish notation that he invented a short time later, he describes his logic, with the notation of the time, based on the definition of identity and implication. He then discusses how this third logical value demolishes some classical tautologies, such as the law of the excluded middle. The motivations for this invention by Lukasiewicz are manifold and the text with which Deaño (in Spanish) precedes its compilation can be consulted (Deaño, in Lukasiewicz 1975) together with the text by Malinowski (1993). However, a short time later, the central motif of this invention became the search for mathematical representations, dependent on veritative variables, for the basic modalities "necessity" and "possibility."

This work of 1920 had wide repercussions in the field of mathematical logic. At the same time, other researchers besides Lukasiewicz created variants and developments concerning how to expand the number of veriative values of logical operations. Here we will limit ourselves to showing the truth tables associated with the three-value logic proposed by Lukasiewicz in 1920, and then we will mention the variant introduced by Kleene in 1938 (Malinowski 1993).

Let us start by defining, for this logic, the set of truth values τ:

$$\tau = \left\{ 1, \; 1/2, \; 0 \right\},$$

where 1 symbolizes the value "true," 1/2 the value "uncertain" or "undecidable," and 0 the value "false." For these trivalent logics, monadic and dyadic operations are also defined according to the applications:

$$\mu : \tau \rightarrow \tau,$$

$$\delta : \tau \times \tau \rightarrow \tau.$$

Below are the tables associated with Lukasiewicz's trivalent logic.

Trivalent negation is defined by

https://doi.org/10.1515/9783112230053-006

p	$\neg p$
1	0
½	½
0	1

As can be seen, it is identified with the classical definition for 1 and 0, but for the intermediate value it happens that $p = \neg p$, which introduces a novelty with respect to classical logics. Kleene's trivalent logic also uses this definition.

Lukasiewicz's table for basic dyadic operations is

$p\ q$	$p \Rightarrow q$	$p \wedge q$	$p \vee q$	$p \equiv q$
1 1	1	1	1	1
1 ½	½	½	1	½
1 0	0	0	1	0
½ 1	1	½	1	½
½ ½	1	½	½	1
½ 0	½	0	½	½
0 1	1	0	1	0
0 ½	1	0	½	½
0 0	1	0	0	1

The Kleene table is almost identical except in its assignment of truth values in cases where uncertain values arise in implication and equivalence. For Kleene's implication, we have:

$$\left(\tfrac{1}{2}, \tfrac{1}{2}\right) \Rightarrow \tfrac{1}{2},$$

and for the equivalence of Kleene, it is

$$\left(\tfrac{1}{2}, \tfrac{1}{2}\right) \equiv \tfrac{1}{2}.$$

In the 1920 text, Lukasiewicz, after defining implication, defines negation as follows:

$$\neg p = (p \Rightarrow 0)$$

and then, armed with implication and negation, he defines the other basic connectives:

$$p \vee q = (p \Rightarrow q) \Rightarrow q$$

$$p \wedge q = \neg\,(\neg\, p \vee \neg\, q)$$

$$p \equiv q = [(p \Rightarrow q) \wedge (q \Rightarrow p)].$$

Sheffer's, Peirce's, and XOR's connectives result from the trivalent negation of conjunction, disjunction, and equivalence, respectively.

The intellectual mechanisms followed by Lukasiewicz to construct his trivalent operations have apparently not been recorded (Urquhart 2001). The reader can perform the exercise of relying on the tables of bivalent logic and deduce what truth value he would assign if there are uncertain variables. This exercise (explicitly carried out in Urquhart's article) involves assuming that uncertain values can potentially be true or false.

But it is important to establish that the objective of these researchers in increasing the number of truth values did not seem to be to model our mental processes, as a contemporary neurobiologist would understand, or as George Boole tried for the case of rational thought, but to extend the theory of formal systems, as we will discuss later. However, some of Lukasiewicz's philosophical considerations, alongside technical motivations, were his strong convictions about cognitive freedom. In a famous 1918 lecture delivered at the University of Warsaw, Lukasiewicz says "I have declared a spiritual war against all coercion that restricts man's free creative activity" (Lukasiewicz 1918). And then Lukasiewicz comments on the two coercions that bind us: physical coercion and logical coercion, continuing this fascinating lecture with comments on his attempts to show that some logical postulates might not be immutable.

At this stage of the development of formal logic, the prevailing objectives were no longer those of the algebra of Boolean logic. Logic was now not an activity to be modeled by mathematics in a similar way to how mathematics modeled celestial mechanics in the time of Laplace and Lagrange, the great inspirers of Boole. In those first decades of the twentieth century, the focus was reversed and an attempt was made to understand the logical foundations of mathematics, a science that from Euclid onward was based on axioms and rules of demonstration based on logic. After the formalization of axiomatic systems by Peano or by Russell and Whitehead, the development of logic took the path of seeking to understand the potential and limitations of these axiom systems. This path prepared the arrival of the great discoveries of Gödel and Turing and the formal developments that led to modern computer science (Davis 2000).

Let us comment that the implication operation whose table we have shown in Chapter 1, so relevant in several of the tautologies that support axiomatic systems and mathematical demonstrations, is the so-called material implication. There is a secular debate as to whether this implication represents the "if . . .. then" of human reasoning. This debate, which began in the time of the Greek logicians, has not ceased. The interested reader can consult any article on the so-called "paradoxes of material implication," for example, statements like: "If every even number is divisible by three, then humans live inside the Sun." turn out to be true. We will not go into this issue where there are plenty of defenders and accusers of material implication. We will only point out that its prevalence in logic is probably due to its operational clarity in a formal system. We have already seen how Lukasiewicz relies on implication to define the other relevant connectives, including negation, and on this he based the creation of his third truth value. Therefore, the exercise of extending binary logic to three variables began based on the fundamental tables shown in Chapter 1.

A territory where many-valued logics have had a strong influence, particularly since the work of Lofti Zadeh in the 1960s, is that of fuzzy sets and fuzzy logics. The interest in these formalisms was born in electrical engineering and control theory. Today, it is an intense field of research and development in technological areas (on the fundamentals, see Klir and Folger (1988)).

The formalism of matrix operators described above is hospitable to expand to more than two truth values. The three variables of the Lukasiewicz and Kleene logics can be directly introduced into the structure of the matrices. We will see this in the next chapter when we analyze one of the ways of representing logical modalities. Here, however, we want to explore another situation: *What happens if a binary logical matrix is confronted with a linear combination of vectors s and n?*

In this linear combination, it is possible to define the scalars so that the resulting vectors can be interpreted as located between s and n, thus representing intermediate veritative values, including the exactly intermediate ones of the trivalent logics of Lukasiewicz and Kleene. Consequently, a procedure emerges here to incorporate uncertainty, not in the operators but in the data. This allows the generation of multipurpose logics of infinite veritative values by means of binary operators and eventually uncertain data. This is the topic that we will see below.

5.1 Weighted vector logic

Suppose that our vector truth values are the Q-dimensional column vectors s and n of the set

$$\tau = \{s, n\}.$$

We now define the set of weighted (or "uncertain" or "fuzzy") vectors as follows (Mizraji 1992):

$$\Pi = \{\gamma s + (1 - \gamma)n: \quad \gamma \in [0, 1]\}. \tag{5.1}$$

Restricting γ to the closed interval [0,1], we weighted n by the complementary of s, so that the coefficients are formally associable with probabilities. This probabilistic weighting has fundamental operative virtues that justify it. Let us point out two of these virtues:

1) At the extremes of the interval $\gamma = 1$ and $\gamma = 0$, the classical truth values are recovered and inside the interval, the value of γ indicates whether the logical variable is closer to the truth or falsehood, with equidistance $\gamma = 1/2$ being the mark of maximum uncertainty.

2) The operations of binary matrices, whether monadic or dyadic, on these vectors generate responses that also have a probabilistic weighting.

Let us demonstrate this second important property. The matrices are linear operators, so that a matrix M operates on a column vector $z = \varphi_1 x + \varphi_2 y$ as follows:

$$M(\varphi_1 x + \varphi_2 y) = \varphi_1 M x + \varphi_2 M y.$$

Let us see then how monadic and dyadic operators behave in the face of probabilized vectors.

5.1.1 Monadic operators

We express any of these operators as

$$U = a s^T + b n^T \; ; \quad a, b \in \tau.$$

Then

$$U[\gamma s + (1 - \gamma) n] = \gamma a + (1 - \gamma) b. \tag{5.2}$$

As we can see, in the general case, the weighting is probabilistic, even when K and M give s and n, respectively, as the final result.

5.1.2 Dyadic operators

Now we express any dyadic operator as the matrix

$$T = a \, (s \otimes s)^T + b \, (s \otimes n)^T + c \, (n \otimes s)^T + d \, (n \otimes n)^T \; ; \quad a, b, c, d \in \tau,$$

and suppose that their inputs are two vectors of the form

$$u = \alpha s + (1 - \alpha) n \text{ and } v = \beta s + (1 - \beta) n.$$

Let us now look at Kronecker's product structure $u \otimes v$:

$$u \otimes v = \alpha \beta \, (s \otimes s) + \alpha(1 - \beta) \, (s \otimes n) + (1 - \alpha)\beta \, (n \otimes s) + (1 - \alpha)(1 - \beta) \, (n \otimes n). \tag{5.3}$$

By the linearity of the matrix T, any of the dyadic connectives will produce an output given by the following equation:

$$T(u \otimes v) = \Phi_1 \, s + \Phi_2 \, n, \tag{5.4}$$

where Φ_1 and Φ_2 result from the grouping and distribution of the coefficients of the terms of eq. (5.3). The sum of the coefficients of eq. (5.3) shows the following:

$$\begin{aligned} \alpha\beta &+ \alpha(1 - \beta) + (1 - \alpha)\beta + (1 - \alpha)(1 - \beta) \\ &= [\alpha + (1 - \alpha)][\beta + (1 - \beta)] = 1 \end{aligned}$$

Therefore, since the terms are non-negative, whatever the order in which they are placed in Φ_1 and Φ_2, it must necessarily be fulfilled $\Phi_1 + \Phi_2 = 1$. Therefore, the resulting vectors are probabilistic, which shows that the set Π is closed to the operations performed by the bivalent logical operators.

5.2 Scalar projections

Since the uncertainty of the veritative vector is represented by a single scalar parameter, we can generate projections that give us explicit scalar equations associated with the different matrix operators. This allows us to clearly understand how the level of uncertainty of the input data influences the uncertainty of the response. We proceed as follows. Given a Out $\in \Pi$, response of a logical operator Ω, we define the scalar projection of that response as

$$\mu_\Omega = s^T \text{Out}. \tag{5.5}$$

This operation retains the Out coefficients that multiply the vector s, implying the complementary coefficient that multiplies n.

We will first analyze the four scalar projections associated with monadic operators for an input of the form $\delta s + (1 - \delta)n$.

M1) *Identity*

$$\mu_I(\delta) = s^T I \left[\delta s + (1 - \delta)n\right] = \delta.$$

M2) *Negation*

$$\mu_N(\delta) = 1 - \delta.$$

M3) *Affirmation*

$$\mu_M(\delta) = 1.$$

M4) *Denial*

$$\mu_K(\delta) = 0.$$

For $\delta = 1$ or $\delta = 0$, the results of these functions are those of the classic tables. But outside of these values appear the novelties of polyvalent logics. In particular, for negation, the value 1/2 produces equality

$$\mu_N(1/2) = 1/2,$$

which transgresses the classic non-truth of $p = \neg p$. Note then that if we restrict δ to the triple 1, 1/2, 0, we get for μ_N the values of the Lukasiewicz and Kleene tables.

Now let us look at the scalar projection associated with dyadic operators, with the input vectors being $u, v \in \Pi$ with the structure $u = \alpha s + (1 - \alpha)n$ and $v = \beta s + (1 - \beta)n$.

D1) *Conjunction*

$$\mu_C(\alpha, \beta) = \alpha\beta.$$

D2) *Disjunction*

$$\mu_D(\alpha, \beta) = \alpha + \beta - \alpha\beta.$$

D3) *Implication*

$$\mu_L(\alpha, \beta) = 1 - \alpha(1 - \beta).$$

D4) *Equivalence*

$$\mu_E(\alpha, \beta) = \alpha\beta + (1 - \alpha)(1 - \beta).$$

D5) *Sheffer's connective*

$$\mu_S(\alpha, \beta) = 1 - \mu_C(\alpha, \beta).$$

D6) *Peirce's connective*

$$\mu_P(\alpha, \beta) = 1 - \mu_D(\alpha, \beta).$$

D7) *Exclusive-or*

$$\mu_X(\alpha, \beta) = 1 - \mu_E(\alpha, \beta).$$

These functions are continuous in $[0, 1]$, but novel situations arise when the two variables have their values in the open interval $(0, 1)$. Note that in all cases where $\alpha, \beta \in \{0, 1\}$ these functions produce the results of the classical bivalent tables. In particular, it is interesting to study the values that these functions take for the various combinations of α and β when we restrict their values to the pairs $(0, 1/2)$, $(1, 1/2)$, and $(1/2, 1/2)$. Let's look at some cases.

D'1) *Conjunction*

$$\mu_C(0, 1/2) = \mu_C(1/2, 0) = 0,$$
$$\mu_C(1, 1/2) = \mu_C(1/2, 1) = 1/2,$$
$$\mu_C(1/2, 1/2) = 1/4.$$

D'2) *Disjunction*

$$\mu_D(0, 1/2) = \mu_D(1/2, 0) = 1/2,$$
$$\mu_D(1, 1/2) = \mu_D(1/2, 1) = 1,$$

$$\mu_D(1/2, 1/2) = 3/4.$$

D'3) *Implication*

$$\mu_L(0, 1/2) = 1, \mu_L(1/2, 0) = 1/2,$$

$$\mu_L(1, 1/2) = 1/2, \mu_L(1/2, 1) = 1,$$

$$\mu_L(1/2, 1/2) = 3/4.$$

D'4) *Equivalence*

$$\mu_E(0, 1/2) = \mu_E(1/2, 0) = 1/2,$$

$$\mu_E(1, 1/2) = \mu_E(1/2, 1) = 1/2,$$

$$\mu_E(1/2, 1/2) = 1/2.$$

The comparison with the trivalent tables of Lukasiewicz and Kleene shows remarkable coincidences. The only discrepancies between the two tables appear for $\alpha = \beta = 1/2$. In the case of equivalence, note that $\alpha = \beta = 1/2$ produce 1/2 as in Kleene's table. *This is especially interesting if we consider that this pseudo-trivalence with results so close to the Lukasiewicz and Kleene tables, is completely generated by bivalent matrix operators.*

In his original paper on trivalent logic, Lukasiewicz defines disjunction as follows:

$$p \vee q = [(p \Rightarrow q) \Rightarrow q],$$

Its matrix version is

$$D'(u, v) = L[L(u \otimes v) \otimes v],$$

and its scalar projection is

$$\mu_{D'}(\alpha, \beta) = 1 - [1 - \alpha(1 - \beta)](1 - \beta).$$

You can see that for the pairs (0, 1/2), (1, 1/2), and (1/2, 1/2), this disjunction reproduces the trivalent tables, except for $\alpha = \beta = 1/2$, where $\mu_{D'}(1/2, 1/2) = 5/8 = 0,625$.

5.3 Tautologies

We will analyze the scalar projections of three of the fundamental tautologies.

5.3.1 Excluded middle

Its classic version states that the formula is always true $p \vee \neg p$. The matrix representation is

$$EM = D(I \otimes N)(u \otimes u) = L(u \otimes u)$$

with the associated projection

$$\mu_{TE}(\alpha) = 1 - \alpha(1 - \alpha).$$

Note that this function has a minimum for $\alpha = 0,75$.

5.3.2 Modus ponens

One of the classic versions is $[p \wedge (p \Rightarrow q)] \Rightarrow q$. The matrix version is

$$MP = L\{C[u \otimes L(u \otimes v)] \otimes v\},$$

and its projection is

$$\mu_{MP} = 1 - \alpha[1 - \alpha(1 - \beta)](1 - \beta).$$

5.3.3 Hypothetical syllogism

If we adopt this version of the hypothetical syllogism

$$[(p \Rightarrow r) \wedge (r \Rightarrow q)] \Rightarrow (p \Rightarrow q).$$

Adding to our u, v pair, the vector $w = \delta s + (1 - \delta)n$, its matrix version is

$$HS = L\{C[L(u \otimes w) \otimes L(w \otimes v)] \otimes L(u \otimes v)\},$$

with the projection

$$\mu_{HS} = 1 - \alpha(1 - \beta)[1 - \alpha(1 - \delta)][1 - \delta(1 - \beta)].$$

5.4 "Shefferian" vector logic

Sheffer's remarkable finding, which allows all classical logical operators to be expressed using a single connective, makes it interesting to express his findings in terms of the Shefferian matrix S. We will look at this for some of the basic logical connectives.

Negation: $\neg p = (p|p)$.

Its matrix representation is

$$N_S(u) = S(u \otimes u) = Su^{[2]}.$$

Its scalar projection being $\mu_{NS} = 1 - \alpha^2$.

Conjunction: $p \wedge q = [(p|q)|(p|q)]$.
Its associated matrix is

$$C_S(u, v) = S[S(u \otimes v) \otimes S(u \otimes v)] = S\left[S^{[2]}(u \otimes v)^{[2]}\right],$$

and its projection is

$$\mu_{CS}(\alpha, \beta) = 1 - (1 - \alpha\beta)^2.$$

Disjunction: $p \wedge q = [(p|p)|(p|q)]$.
Its matrix representation is

$$D_S(u, v) = S[S(u \otimes u) \otimes S(v \otimes v)] = S\left[S^{[2]}\left(u^{[2]} \otimes v^{[2]}\right)\right]$$

with the projection

$$\mu_{DS}(\alpha, \beta) = 1 - \left(1 - \alpha^2\right)\left(1 - \beta^2\right).$$

Implication: $p \Rightarrow q = [p|(q|q)]$.
Its corresponding matrix is

$$L_S(u, v) = S[u \otimes S(v \otimes v)] = S\left(u \otimes Sv^{[2]}\right)$$

with

$$\mu_{LS}(\alpha, \beta) = 1 - \alpha\left(1 - \beta^2\right).$$

There are at least two remarkable aspects in this representation:
1) If in the scalar projections exponents 2 shifted so that $2 \to 1$, the resulting equations are the same basic scalar projections shown above.
2) It does not hold for $\alpha \in (0, 1)$, the double negative rule: $\neg\neg p = p$, except for a fixed point whose value is

$$\alpha^* = \frac{1}{2}\left(\sqrt{5} - 1\right) \approx 0,618 \ldots$$

which is the reciprocal of the golden ratio.

References

Davis, M. (2000) The Universal Computer: The Road from Leibniz to Turing, Norton, New York.

Deaño, A. (1975) Presentación, in J. Lukasiewicz, Estudios de Lógica y Filosofía (A. Deaño, Ed), Revista de Occidente, Madrid.

Klir, G.J., Folger, T.A. (1988) Fuzzy Sets, Uncertainty, and Information, Prentice-Hall, New Jersey.

Lukasiewicz, J. (1918) Farewell lecture by professor jan lukasiewicz, Delivered in the Warsaw university lecture hall on march 7, in Reprinted in J. Lukasiewicz, Selected Works, pp. 153–178, (L. Borkowski, Ed), North-Holland, 1980, Amsterdam.

Lukasiewicz, J. (1920) On three-valued logic, in Reprinted in J. Lukasiewicz, Selected Works, pp. 87–88, (L. Borkowski, Ed), North-Holland, 1980, Amsterdam.

Malinowski, G. (1993) Many-Valued Logics, Clarendon Press, Oxford.

Mizraji, E. (1992) Vector logic: The matrix-vector representation of logical calculua, Fuzzy Sets and Systems, 50: 179–185.

Urquhart, A. (2001) Basic many-valued logic. in Handbook of Philosophical Logic, pp. 249–295, (D.M. Gabbay, F. Guenthner, Eds), Kluwer, Dordrecht.

Chapter 6
The modalities and conceptualization of uncertainty

Chapter XLIV of Edward Gibbon's "Decline and Fall of the Roman Empire" begins with these majestic sentences: "The vain titles of the victories of Justinian are crumbled into dust; but the name of the legislator is inscribed on a fair and everlasting monument. Under his reign, and by his care, the civil jurisprudence was digested in the immortal works of the Code, the Pandects, and the Institutes." In imitation of Gibbon, we could say of Aristotle that his physics and astronomy lie on the ground, swept away by history, but that his treatises on logic subsist as a grandiose and everlasting monument. In using his words and referring to Aristotle, we do not believe that we contradict Gibbon's appreciation of the Greek philosopher. In the same chapter dedicated to Roman jurisprudence, referring to the work of the jurist Servius Sulpicius, Gibbon writes: "[. . .] Servius Sulpicius was the first civilian who established his art on a certain and general theory. For the discernment of truth and falsehood he applied, as an infallible rule, the logic of Aristotle and the stoics, reduced particular cases to general principles, and diffused over the shapeless mass the light of order and eloquence." Aristotle's logic has been mistreated and then reviled by the misuses made of it centuries later by scholastics and dogmatic academics. But today, well into the twenty-first century, it is still possible to extract poignant and inspiring observations from his treatises.

Of Aristotle's works of logic, it is in the book *On Interpretation* (circa 350 BC) that the fundamental theorems of modal logic appear. These theorems were expressed verbally, but as we will see in a famous example, their statements were already well prepared to receive a mathematical representation that had to wait until the first decades of the twentieth century. *On Interpretation* is a book at once remarkably short and extremely rich on ideas. It begins by characterizing the basic components of human language, nouns, and verbs. Then he addresses the construction of propositions and the attribution to them of the veritative values "true" and "false." It is in chapter 9 of that work that Aristotle makes his now famous observations on contingent futures and the impossibility of some well-constructed propositions being judged as true or false. "Tomorrow there will be a naval combat," is the always cited example that prefigures the existence of undecidable propositions and of a third value of truth assignable to the uncertain. Today, the existence tomorrow of that naval combat is neither true nor false. It is significant that Lukasiewicz, who in his 1920 work described his trivalent logic as non-Aristotelian, has abandoned this denomination in later works and has cited in the appendix of his central work on modalities, those reflections of Aristotle on contingent futures (Lukasiewicz 1930).

https://doi.org/10.1515/9783112230053-007

Modalities are usual forms of language. Here are two examples: "it is possible that someone is now reading the history of the decline of the Roman Empire" and "in order to write this sentence it is necessary that I know the letters of this language I use." "Possible" and "necessary" are the two basic modalities whose subtle meaning was established by Greek logic and exhaustively analyzed by Aristotle in *On Interpretation*. This initial link between modalities and thought brings us back here to the idea, prior to the logicism of the twentieth century, which conceived of logic as the science associated with the rules of correct reasoning.

To exemplify the way in which Aristotle expounded verbal theorems, let's see how he enunciates the link between possibility and necessity. He does so in chapter 13 of *On Interpretation*, where he writes: "It remains therefore that only the proposition 'it is not necessary that it is not' follows the proposition 'it is possible that it is'" (translated from the French edition by J. Tricot, which is the one we cited in the references).

In the following sections, we will present two ways to construct matrix models for logical modalities. In the first, the modalities are constructed by means of matrices of bivalent logic that act on potentially uncertain logical vectors. In the second, we will incorporate Lukasiewicz and Tarski's results into the structure of the logical matrices themselves, expanding the set of truth vectors. For the reasons we will see, we will refer first to "current modalities" and then to "conceptualized uncertainty."

6.1 Current modalities

As with sets, logical operations can be defined by extension or by comprehension. Truth tables are definitions by extension. But operations that are clear to human cognition in its linguistic coding, such as necessity or possibility, cannot be expressed extensively in binary logic. The impossibility was demonstrated by Lukasiewicz and one of his students, Alfred Tarski, found a definition by comprehension within the framework of trivalent logic, to which a truth table could then be assigned.

The usual symbols to represent the basic modalities (there are others that we will deal with later) are $\Diamond(p)$ for the proposition "p is possible" and $\Box(p)$ for "p is necessary." In this symbolism, Aristotle's theorem is represented by the equation

$$\Diamond(p) = \neg\,\Box\,(\neg p). \tag{6.1}$$

Lukasiewicz's demonstration of the impossibility of representing these modal connectives within bivalent logic arises from the following (reasonable and intuitive) evaluation of these two modalities for the values true, 1, and false, 0:

$$\Diamond(1) = 1\,;\quad \Diamond(0) = 0;$$

$$\Box\,(1) = 1\ ;\quad \Box(0) = 0.$$

Consequently, for these binary variables, the operators cannot be distinguished. Another argument is that both modalities are monadic operators, but none of the four monadic operators, which we have discussed in another chapter, can represent what these modalities mean.

Tarski's proposal, within the framework of Lukasiewicz's trivalent logic, is to define the modalities by comprehension by means of the following logical formula:

$$\Diamond(p) = \neg p \Rightarrow p. \tag{6.2}$$

If this operation is evaluated for the three values of the trivalent logic, it results:

$$\Diamond(1) = 1; \quad \Diamond(1/2) = 1; \quad \Diamond(0) = 0,$$

which is intuitively acceptable because the uncertain is possible. Using Aristotle's theorem, it results in:

$$\Box(p) = \neg\Diamond(\neg p) = \neg(p \Rightarrow \neg p), \tag{6.3}$$

and its evaluation produces

$$\Box(1) = 1; \quad \Box(1/2) = 0; \quad \Box(0) = 0,$$

a result also consistent with what was expected of a good definition. In this way, trivalent logics, along with their philosophical meanings, here showed a real technical potential to provide modal logic with acceptable definitions, both through closed formulas and truth tables.

If we now represent Tarski's formula by means of the matrix operator formalism of two-valued logic, and assuming as inputs vectors with probabilistic weights, we obtain

$$\Diamond[u] = L(Nu \otimes u), \quad u = \alpha s + (1-\alpha)n \in \Pi. \tag{6.4}$$

If we evaluate this expression for $\alpha = 1, \alpha = 1/2, \alpha = 0$, it turns out that

$$\Diamond[s] = s; \quad \Diamond[(1/2)s + (1/2)n] = (3/4)s + (1/4)n; \quad \Diamond[n] = n.$$

Here, we see that the representation fails for the value of u which represents the maximum uncertainty.

However, we will show below that Tarski's representation suggests a recursion that leads to a satisfactory construction of modalities (Mizraji 1994). In addition, the result of this process produces modalities that expand its evaluation for uncertainties spread over the continuous interval $[0, 1]$. In the recursive process that we will show, the logical matrices are bivalent, the vectors belong to the set Π of probabilized vectors and modality evaluations are updated at each step of the recursion. Due to this update we have named this section "current modalities." Let us start by noting that Tarski's operation can be transformed into a disjunction:

$$L(Nu \otimes u) = L(N \otimes I)(u \otimes u) = D(u \otimes u),$$

with

$$u = as + (1 - \alpha)n.$$

From this result, we establish the following recursive process:

$$u_1 = D(u \otimes u)$$
$$u_2 = D(u_1 \otimes u_1)$$
$$\cdots$$
$$u_{r+1} = D(u_r \otimes u_r).$$

Let us define the operation as "possibility of order r"

$$\Diamond_r [u] = D(u_r \otimes u_r) . \tag{6.5}$$

Finally, we define the possibility of u as the limit

$$\Diamond [u] = \lim_{r \to \infty} \Diamond_r [u]. \tag{6.6}$$

Regarding the operation "necessity," we begin by showing that in this representation, Aristotle's theorem (6.1) can be considered (reversing the sense of historical time) as a corollary of De Morgan's law. We write eq. (6.1) as follows:

$$\Box (q) = \neg \Diamond (\neg q).$$

This expression is proven from (6.1) by negating $\Diamond(p)$ and defining $q = \neg p$. Returning to the matrix representation and considering De Morgan's theorem $C = ND(N \otimes N)$, then we have the following matrix expression for the necessity:

$$\Box [v] = N \Diamond [Nv] = ND(Nv \otimes Nv) = C(v \otimes v). \tag{6.7}$$

Let us take $v = \beta s + (1 - \beta)n \in \Pi$. We see that this Eq. (6.7) also does not satisfy the conditions of a "necessity" operator for intermediate truth values. Now, we build the type of iteration we perform for the operator possibility:

$$v_1 = C(v \otimes v)$$
$$v_2 = C(v_1 \otimes v_1)$$
$$\cdots$$
$$v_{r+1} = C(v_r \otimes v_r).$$

Let us define the "necessity of order r" as the operation

$$\Box_r [v] = C(v_r \otimes v_r). \tag{6.8}$$

and necessity is also defined as the limit of the sequence

$$\Box [v] = \lim_{r \to \infty} \Box_r [v]. \tag{6.9}$$

The modalities of order r correspond to matrix representations

$$\Diamond_r[u] = \left(\prod_{i=0}^{r-1} D^{[f(i)]}\right) u^{[f(r)]},$$ (6.10)

$$\Box_r[v] = \left(\prod_{i=0}^{r-1} C^{[f(i)]}\right) v^{[f(r)]}.$$ (6.11)

where $f(z) = 2^z$ $(z = 1, \ldots, r)$.

At the limit, the modalities are defined by matrices that can be represented by

$$\Diamond [u] = \left(\prod_{i=0}^{\infty} D^{[f(i)]}\right) u^{[\infty]},$$ (6.12)

$$\Box [v] = \left(\prod_{i=0}^{\infty} C^{[f(i)]}\right) v^{[\infty]}.$$ (6.13)

Note that in these unusual operations, matrices of order $Q \times$ Infinity act on infinite column vectors.

Despite this apparent complexity, the fact that the set of probabilistic vectors Π is closed to the operations of the logical matrices causes each step of the iterations to produce vectors in Π whose coefficients can be calculated. This fact allows equations for the limits of the sequences to be obtained explicitly from the calculation of the scalar coefficients.

It is easy to show that the scalar projections of the r-order modalities are given by

$$\Diamond_r[a] = s^T \Diamond_r[u] = 1 - (1 - \alpha)^t, \quad t = 2^r,$$ (6.14)

$$\Box_r[\beta] = s^T \Box_r[v] = \beta^t, \quad t = 2^r.$$ (6.15)

From these expressions, it emerges that the limits for $r \to \infty$ are

$$\Diamond(\alpha) = \begin{cases} 1 & \text{if} \quad \alpha \neq 0 \\ 0 & \text{if} \quad \alpha = 0 \end{cases},$$ (6.16)

$$\Box(\beta) = \begin{cases} 1 & \textit{if} \quad \beta = 1 \\ 0 & \textit{if} \quad \beta \neq 1 \end{cases}.$$ (6.17)

These projections unequivocally define the boundary vectors. The most interesting (and perhaps important) consequences of the previous analysis are the following: (a) These modalities are generated by *bivalent logical matrices* acting on probabilistic vectors; (b) the evaluation of possibility and necessity for certainty (1 and 0) and uncertainty are applied to the continuous interval [0,1], which is an extension of the representation of modalities by trivalent logic, and which includes it.

6.1.1 Extensive quantification

There is a strong formal analogy between the possibility modality $\diamondsuit$ and the existential quantifier $\exists$, on the one hand, and between necessity $\square$ and the universal quantifier $\forall$ on the other. Both quantifiers are linked by a relationship formally identical to Aristotle's theorem of modalities. Let us exemplify this:

The statement "the function of real variables $y = x^2$ is such that $\forall x$ it is $y \geq 0$" is equivalent to stating "given the function of real variables $y = x^2$ does not $\exists x$ such that it is $y < 0$." Note that it $y < 0$ is the negation of $y \geq 0$. If we generalize this structure as is done in the calculus of first-order predicates, and our mathematical statement regarding inequality is represented by a predicate $P(x)$, we will say $\forall x \, P(x)$ and see that this is equivalent to affirming $\neg \, \exists \, x \, [\neg \, P(x)]$.

Suppose that $P(x)$ is a predicate and that there exists a universe defined by a finite set of variables x_i:

$$U = \{x_1, x_2, \ldots, x_R\}.$$

From this set, quantifiers can be symbolically characterized by the following definitions by extension

$$\exists \, P(x) = P(x_1) \vee P(x_2) \vee \cdots \vee P(x_R), \tag{6.18}$$

$$\forall \, P(x) = P(x_1) \wedge P(x_2) \wedge \cdots \wedge P(x_R). \tag{6.19}$$

Naturally, the binary character of the disjunction and the conjunction does not allow these concatenations to be processed, but what is symbolized here are operations that must be performed recursively, including, at each step, new elements of the set U.

Suppose that each predicate $P(x)$ is identified with the attribution of a vector truth value. Let us now see how the previous expressions guide the construction of the vector versions of the existential and universal quantifiers for probabilistic vectors of the form $u_i = \alpha_i s + (1 - \alpha_i)n$ and $v_i = \beta_i s + (1 - \beta_i)n$. We establish the following recursions:

$$d_1 = u_1$$
$$d_2 = D(u_2 \otimes d_1)$$
$$\cdots$$
$$d_R = D(u_R \otimes d_{R-1}).$$

We then define the vector version for the existential quantifier over a finite set of R propositional dictates as follows:

$$\exists_R[u] = d_R. \tag{6.20}$$

Using an analogous procedure for the universal quantifier, its vector version is

$$c_1 = v_1$$
$$c_2 = C(v_2 \otimes c_1)$$
$$\cdots$$
$$c_R = C(v_R \otimes c_{R-1}).$$

The vector version of the universal quantifier is

$$\forall_R[v] = c_R. \tag{6.21}$$

The scalar projections associated with these quantifiers are as follows:

$$s^T \exists_R[u] = 1 - (1 - \alpha_1)(1 - \alpha_2) \cdots (1 - \alpha_R), \tag{6.22}$$

$$s^T \forall_R[v] = \beta_1 \, \beta_2 \cdots \beta_R. \tag{6.23}$$

Let us mention, in passing, a curious background: these eqs. (6.22) and (6.23) appear in a footnote to John Maynard Keynes's treatise on probability, where he mentions as a source an article on the credibility of testimonies published anonymously in 1699 (see Keynes 1921 and Anonymous 1669).

Note that if $R \to \infty$, these projections of the quantifiers can be represented by equations identical to (6.16) and (6.17) if the α and β are interpreted to be associated with geometric means (see Mizraji and Lin 1997). But the most important thing to ratify the formal similarities between modalities and quantifiers is that Aristotle's theorem, in its version for quantifiers, can be demonstrated directly by appealing to De Morgan's law, based on the sequences that we have shown.

This similarity leads us to suggest that in some cases, the cognitive evaluation of modalities may be better modeled by quantifier theory than by logical modality theory (Mizraji and Lin 1997, 2001). To exemplify this, let us imagine the following. Someone asks a group of people to answer this question: "Is it possible for any university in the academically developed world to award a Ph.D. in science to a scientist who has not completed all the courses in a university curriculum?" Faced with questions of this type, which do not concern any ideological "should be" but rather data of reality, the people in the group must review their information, whether it is installed in their memories, or eventually, what they obtain from documents. Let us note that the opinion regarding this possibility arises from a tour of information bases, as occurs in the construction model of the existential quantifier that we have just seen.

6.2 Conceptualized uncertainty

Now we will see how there is a way to model the structure of the modalities by appealing to matrix logical operators, which complements the approach we showed in the previous section. Suppose you ask a believer, an agnostic, and an atheist the following: Is it necessary or possible for God to exist? The believer answers "yes, it is necessary," the agnostic says "yes, it is possible," and the atheist responds "no, it is neither necessary nor possible." For those people of well-established opinions, the attribution of truthful values to modalities does not require going through any information base. This is a different situation from the one shown in the previous section.

In the case of the example we have just seen, the modalities "possible" and "necessary," as well as the values "true," "false," and "uncertain" (or "undecidable") are previously conceptualized. But the conceptualized uncertainty is explicit in the truth tables of trivalent logics, which suggests the interest of extending the algebra of matrix logical operators to a set that includes three vectors as truth values.

Let us define the orthonormal set of trivalent truth values as consisting of three Q-dimensional column vectors:

$$\tau_3 = \{s, \quad n, \quad h\},$$

where the vector h is assigned to undecidable propositions. Ortho-normality results in:

$$\langle s, s \rangle = \langle n, n \rangle = \langle h, h \rangle = 1 \text{ and } \langle s, n \rangle = \langle s, h \rangle = \langle n, s \rangle = \langle n, h \rangle = \langle h, s \rangle = \langle h, n \rangle = 0.$$

In this framework, as in the case of bivalent logics, we also define monadic and dyadic matrix operators based on the following applications:

$$\mu_3 : \tau_3 \longrightarrow \tau_3,$$

$$\delta_3 : \tau_3 \times \tau_3 \longrightarrow \tau_3.$$

Unlike what happened with bivalent logic operators, in this case we are facing a situation of combinatorial explosion. The total number of n-adic functions of m variables is given by the formula

$$m^{(m^n)}.$$

The exponent m^n indicates the number of variations with repetition of m variables in groups of n elements. This exponent gives us the number of terms of the logical expression, hence it follows that the total number of functions is m raised to the number of terms of the expression. Let us then note the following.

Bivalent logic
Number of monadic functions: $2^{(2^1)} = 4$
Number of dyadic functions: $2^{(2^2)} = 16$

Trivalent logic
Number of monadic functions: $3^{(3^1)} = 27$
Number of dyadic functions: $3^{(3^2)} = 3^9 = 19.683$

Here, we will start by building the matrix operators associated with the basic logical operations.

6.2.1 Basic monadic operators

6.2.1.1 Identity

$$I_3 = ss^T + nn^T + hh^T.$$

Note that if $z \in \tau_3$ then $I_3 z = z$.

6.2.1.2 Negation

$$N_3 = ns^T + sn^T + hh^T,$$

where $N_3 s = n$, $N_3 n = s$ and $N_3 h = h$, which represents a matrix version of the trivalent negation table.

It can be proven with this negation matrix that the double negative generates an identity:

$$(N_3)^2 = I_3.$$

These formulas can also be expressed using the bivalent matrices (since in both the bivalent and trivalent cases the dimension is $Q \times Q$):

$$I_3 = I + hh^T \quad \text{and} \quad N_3 = N + hh^T.$$

6.2.2 Basic dyadic operators

We will adopt the following general representation for dyadic connectives:

$$Op_3 = a_1(s \otimes s)^T + a_2(s \otimes n)^T + a_3(n \otimes s)^T + a_4(n \otimes n)^T + \\ b_1(s \otimes h)^T + b_2(h \otimes s)^T + b_3(n \otimes h)^T + b_4(h \otimes n)^T + b_5(h \otimes h)^T. \tag{6.24}$$

where $a_i, b_i \in \tau_3$.

Imitating the format usually adopted for complex numbers, we will compact the representation of the operators with format (6.24) by grouping the coefficients as follows:

$$Op_3 = \text{Re}(a_1 a_2 a_3 a_4) + \text{In}(b_1 b_2 b_3 b_4 b_5). \tag{6.25}$$

As can be seen, the term "Re" gives the coefficients of the terms that do not contain the uncertain value h, while the term "In" groups the coefficients of the uncertain terms.

We begin by basing ourselves on conjunction and disjunction.

6.2.2.1 Conjunction

$$C_3 = \mathrm{Re}(snnn) + \mathrm{In}(hhnnh)$$

If the expanded version is used, it is seen that the operations on vectors τ_3 reproduce the trivalent table of the conjunction, as is expected because the matrix is a mere translation into linear algebra of that table:

$$C_3(s \otimes s) = s; \; C_3(s \otimes n) = C_3(n \otimes s) = C_3(n \otimes n) = n;$$

$$C_3(n \otimes h) = C_3(h \otimes n) = n; \; C_3(s \otimes h) = C_3(h \otimes s) = C_3(h \otimes h) = h.$$

6.2.2.2 Disjunction

$$D_3 = \mathrm{Re}(sssn) + \mathrm{In}(sshhh)$$

From these definitions, it can be easily calculated that these trivalent matrices comply with De Morgan's laws:

$$C_3 = N_3 D_3 (N_3 \otimes N_3), D_3 = N_3 C_3 (N_3 \otimes N_3).$$

At the same time, these definitions allow us to define the trivalent versions of Sheffer and Peirce's connectives:

$$S_3 = N_3 C_3 = \mathrm{Re}(nsss) + \mathrm{In}(hhssh),$$

$$P_3 = N_3 D_3 = \mathrm{Re}(nnns) + \mathrm{In}(nnhhh).$$

6.2.2.3 Implication

Let us now see what results from the attempt to define implication using one of the classical definitions in bivalent logic

$$L_3 = D_3 (N_3 \otimes I_3).$$

If the operations are performed, the result is

$$L_3 = \mathrm{Re}(snss) + \mathrm{In}(hsshh).$$

This matrix implication reproduces, in vector version, exactly the table of the Kleene implication. This shows that here, the matrix formalism for trivalent logics introduces divergences with Lukasiewicz's logic, despite having as its starting point the negation, conjunction and disjunction that Lukasiewicz defined. This divergence will have consequences on the structure of operations that are defined using implication, such as equivalence and or-exclusive.

If we impose Lukasiewicz's implication as a primitive concept and not derived from theorems, we have the matrix

$$L_{3P} = \mathrm{Re}(snss) + \mathrm{In}(hsshs),$$

where we add the subscript P (alluding to Polish logic) that we will also include in the functions that we derive from this structured matrix implication.

6.2.2.4 Equivalence and exclusive-or

The definitions of equivalence and exclusive-or offer two types of problems:

1. If we derive the equivalence of the classical theorem

$$p \equiv q = [(p \Rightarrow q) \wedge (q \Rightarrow p)].$$

This definition will give different results depending on which implication we use.

2. In the matrix version, expressions with this structure will appear

$$\text{EQUI}_{3(P)}(u, v) = C_3 \left[L_{3(P)}(u \otimes v) \otimes \left[L_{3(P)}(v \otimes u) \right] \right], \quad u, v \in \tau_3,$$

which are matrices of order $Q \times Q^4$. The subscript $3(P)$ indicates that this expression – not its result – is structurally the same for basic and Polish three-valued implication. The procedure we have followed is to perform the calculations and associate the set of nine outputs produced with the terms of the general equation (6.24) (Mizraji 2008). In this way we force the reduction of the dimension of the equivalence matrix by means of the projections:

$$\text{EQUI}_3(u, v) \rightarrow E_3(u \otimes v),$$

$$\text{EQUI}_{3P}(u, v) \rightarrow E_{3P}(u \otimes v).$$

The results are as follows:

$$E_3 = \text{Re}(snns) + \text{In}(hhhhh),$$

$$E_{3P} = \text{Re}(snns) + \text{In}(hhhhs).$$

The associated exclusive-or are

$$X_3 = N_3 E_3 = \text{Re}(nssn) + \text{In}(hhhhh),$$

$$X_{3P} = N_3 E_{3P} = \text{Re}(nssn) + \text{In}(hhhhn).$$

If Polish disjunction is defined using the formula

$$D_{3P} = L_{3P}(N_3 \otimes I_3).$$

we obtain a disjunction that differs from the one originally defined by the tables of trivalent logic:

$$D_{3P} = \text{Re}(sssn) + \text{In}(sshhs).$$

From De Morgan's law results the matrix for the "Polish" conjunction, also different from the one presented at the beginning

$$C_{3P} = \text{Re}(snnn) + \text{In}(hhnnn).$$

To complete these Polish connectives, we can immediately derive the Polish functions of Sheffer and Peirce operators:

$$S_{3P} = N_3 C_{3P} = \text{Re}(nsss) + \text{In}(hhsss),$$

$$P_{3P} = N_3 D_{3P} = \text{Re}(nnns) + \text{In}(nnhhn).$$

As can be seen, this matrix formalism allows us to define from the Polish implication (which is singled out by expressing that $(1/2) \Rightarrow (1/2)$ it has as output the value 1), an internally consistent system that alters the structure of the basic connectives described by some of the original tables derived from Lukasiewicz's logic. This indicates that there are some potentially interesting formal topics, which we have not explored further.

6.2.3 Trivalent modalities

Within the framework of this trivalent vector logic, we will designate the matrices that define the modalities, possibility, and necessity Pos and Nec, respectively. These are monadic matrices that are within the set of the 27 possible monadic matrices in a trivalent logic. At the beginning of this chapter we saw that the incorporation of the third truth value 1/2 made it possible to obtain explicit representations for possibility and necessity. In the matrix format these representations can be directly transcribed from eqs. 6.2 and 6.3, resulting in

$$\text{Pos} = ss^T + nn^T + sh^T,$$

$$\text{Nec} = ss^T + nn^T + nh^T.$$

Consequently,

$$\text{Pos}(s) = \text{Pos}(h) = s, \text{Pos}(n) = n,$$
$$\text{Nec}(s) = s, \text{Nec}(n) = \text{Nec}(h) = n.$$

Note also that these modalities can be expressed using the identity matrix of bivalent logic:

$$\text{Pos} = I + sh^T, \text{Nec} = I + nh^T.$$

These modal matrices satisfy Aristotle's theorem:

$$\text{Pos} = N_3 \text{ Nec } N_3.$$

Part 3: **Explorations of complexity**

Chapter 7
Elements of differential and integral calculus of logical operators

In 1872, Karl Weierstrass reported his famous function

$$W(x) = \sum_{n=0}^{\infty} a^n \cos(b^n \pi \, x),$$

which for certain values of the parameters a and b is continuous but lacks a derivative at all points of the real line. This discovery was a milestone in the history of mathematics. Until that time, it was well understood that a function could be continuous
and present some breaking points without derivatives. But Weierstrass's finding was
radical: continuous along the entire line and without derivative along the entire line.
There the relationship between continuity and differentiability was completely clarified. The existence of derivatives had continuity as a necessary condition, but this
was not a sufficient condition. Thus, the contraposition of "If f is derivable, then f is
continuous" is "If f is not continuous, then f is not derivable." This conclusion seemed
to take away any validity from arguments such as the one used by Boole to obtain
series expansions of logical functions by appealing to MacLaurin's developments, as
we saw in Chapter 1.

However, these well-established results for real variable functions were not an
obstacle for the territory of discrete mathematics to begin to open up new avenues of
exploration. One of the starting points for these explorations was a master's thesis
presented to the Massachusetts Institute of Technology (MIT) by a 21-year-old named
Claude Shannon. The title of this thesis is "A Symbolic Analysis of Relay and Switching
Circuits" (Shannon 1940). The facsimile version of this thesis is available on some web-
sites. A compact version of the argument was published, with the same title, in an
electrical engineering journal (Shannon 1938). H.H. Goldstime, in a book on the evolution of computing, points out that this has been one of the most important master's
theses ever written and adds that it's a benchmark that helped transform digital circuit design from an art to a science. In this foundational work, Shannon discovers
that in the study of the properties of circuits with relays and switches, the algebra of
logic is a completely adapted instrument for the analysis and design of these circuits.
From that moment on, there was a noticeable increase in power in the field of electrical engineering as a result of the application of algebraic logic. In this way, that science born in the nineteenth century and which was somewhat displaced by the logicist paradigm, became a protagonist of the first order from the unexpected angle of
technologies.

Shannon refers in his thesis to two works on the algebra of logic, ancient and fundamental. One is Alfred North Whitehead's erudite and comprehensive treatise on

https://doi.org/10.1515/9783112230053-008

universal algebra (Whitehead 1898), the other is Louis Couturat's short and elegant book (Couturat 1914) in its English translation (he also cites George Boole's treatise on equations of difference published in 1860). In the books of Couturat and Whitehead, the expansion of logical functions created by Boole, and which Shannon uses in his analysis of circuits, is described in detail (an expansion sometimes referred to in some texts as Shannon's expansion, although Shannon never indicated that this invention was his own).

The usefulness that algebraic logic began to show in electrical engineering and then in computer science led to the use of canonical representations of logical operations, where they were described by a syntax based only on conjunctions, disjunctions, and negations. One of the most commonly used canonical representations is the so-called disjunctive normal form, where expressions are organized as packages of disjunctions, as shown in the following example:

$$f(x,y,z) = (x \wedge \neg y) \vee (x \wedge z) \vee (\neg y \wedge z), \qquad x,y,z \in \{0,1\}.$$

Another alternative, also shown by Shannon in his thesis, is to use normal conjunctive forms, where packages of expressions based on disjunctions or negations are separated by conjunctions.

But the powerful mathematical procedures for representing circuits with relays and switches led to an important problem: How to evaluate the effect of a change in a variable of the circuit on the various properties of that circuit? Does the elimination or inclusion of a resistance in one of its sectors change the circuit's overall response to its inputs? It is interesting to note that in this area of engineering, once problems are formalized in terms of discrete variables, the type of problem arises that centuries earlier led Leibniz to invent his version of differential calculus. This problem can be summed up in one question: how to measure change? (for details, see Mizraji (2013)). In the case of circuits represented by logical variables, the answer to this question was equivalent to that found by Leibniz for continuous variable functions. It consisted of inventing a Boolean derivative capable of measuring the change of a function when the variable whose effect is explored changes from 1 to 0 or to 0 to 1. The Boolean derivative of a logical function, usually a partial derivative, is defined as follows:

$$\frac{\partial f(x_1, \ldots, x_i, \ldots, x_n)}{\partial x_i} = \text{XOR}[f(x_1, \ldots, 1, \ldots, x_n), \quad f(x_1, \ldots, 0, \ldots, x_n)],$$

where XOR represents the exclusive-or (Davio, Deschamps, and Thayse 1978; Gorbatov 1988).

Another field of research that led to the use of Boolean derivatives was the theory of cellular automata. These objects are dynamic systems that "inhabit" a space made up of cells. In the simplest cases, these cells can adopt the values 0 and 1 of binary mathematics, and their transformation laws can be expressed by Boolean logical functions. Unlike circuit theory, where the symbolism of logic produced at the same time the possibility of capturing and controlling the complexity of systems, in the field of cellular automata the roots of the complexity remain partially understood. We will devote our last chapters to this topic. Gérard Vichniac (1990) published a refined and exploratory work on the properties of Boolean derivatives, and their use in the theory of cellular automata, in a physics journal.

The development of a differential calculus of Boolean functions induced the search and development of an integral calculus for Boolean functions (Tucker, Tapia, and Bennett 1988). In these theories, which involve Boolean derivatives and integrals, researchers generally operate on canonical representations. These representations, like the disjunctive forms shown above, could be assimilated to the most basic syntax of the representation of logical functions. This could be metaphorically assimilated to low-level programming languages. High-level languages adopt a strategy of grouping elementary operations that shows a kind of intrinsic semantics at that level. Based on this idea, the search for a differential and integral calculus of logical operations based on semantics of a higher level than canonical representations can reveal structures that are not perceived in the usual binary calculus.

The semantics to which we refer can be easily obtained from the matrix structure that we have presented in the previous chapters. We have carried out this exploration (Mizraji 2015) and here we will present some of its results. Thus, we can very easily show that the derivative of one of the classical tautologies, instead of giving the expected 0 of the standard Boolean derivative (since the tautology is invariably equal to 1 for any Boolean value of its variables), produces a logical formula endowed with structure. For example, deriving the *modus ponens* we get

$$\frac{\partial}{\partial p}\{[p\wedge(p\Rightarrow q)]\Rightarrow q\}=\neg\,(q\vee\neg\,q),$$

a result that is 0 in the Boolean domain, but which here shows that it is the negation of another tautology, the law of the excluded middle.

Integral calculus on matrix operators generates interesting results. The first is that integrals are always calculable and expressible by simple equations. This sets a big difference with the integral calculus of real functions, where the indefinite integral of certain functions, even if it exists, may not be expressible with a finite number of terms or factors (as in the famous case of $y(x)=\exp(-x^2)$). In the formalism we present, for a logical function, we can first define a general integral. We illustrate this for a simple case:

$$\int(p\Rightarrow\neg\,q)\,\partial\,r=[(p\Rightarrow\neg\,q)\Rightarrow r].$$

But also, as we shall see, for this same logical function a particular integral can be found:

$$\int_P (p \Rightarrow \neg q)\, \partial r = [(p \vee r) \Rightarrow \neg (q \vee r)].$$

In these equations, ∂r, it is a symbolic differential that indicates the variable with respect to which the logical expression is integrated. Why integrate with respect to a new variable? The answer arises from the fact that one of the characteristics of the Boolean derivative is that it makes the variable with respect to which it is derived disappear, and therefore the antiderivative must be made with respect to a variable that is not in the original expression.

These equations that have just been shown exemplify the research strategy described by mathematician Z.A. Melzak in his book "Bypasses: A Simple Approach to Complexity" (Melzak 1983). The bypass to which the author refers consists of transferring the complex problem to an alternative representation in which the operations are simpler, and once the solutions are obtained, returning to the initial formalism to express this solution in its original framework. The previous results were obtained by translating the logical expressions into the format of matrix logical operators, and operating within the rules of matrix algebra and Kronecker's products.

Let us also point out two new aspects. One concerns the fact that the derivatives and integrals obtained on vectors allow the results to be extended to probabilized vectors such as those analyzed in Chapter 5. The other aspect is related to the loss and gain of information that occur in the calculation of continuous functions when deriving and integrating. In this situation, the derivation usually causes information to be lost by eliminating variables and parameters. Integration, on the other hand, must inject information by appealing to a reduced set of basic integrals and theorems (such as integration by substitution or integration by parts) that allow information present in the basic integrals to be managed. In the differential and integral calculus of logical variables, the two procedures are algorithmic and the loss and gain of information is measured by the reduction and increase of the number of logical variables when deriving or integrating, respectively.

7.1 Derivatives from matrix logic operations

Let us imagine a logical operation represented by matrices, vectors, together with matrix and Kronecker products, and let us symbolize the dependence of a vector u by

$$Op(u) = \Psi(u, v, \ldots, z), \qquad u, v, \ldots, z \in \Pi, \tag{7.1}$$

with $\Pi = \{\gamma s + (1 - \gamma)n: \ \gamma \in [0,1]\}$.

We define the partial derivative with respect to u, assuming the other variables are constant, as follows:

$$\frac{\partial Op(u)}{\partial u} = X[Op(s) \otimes Op(n)] \tag{7.2}$$

where X is the matrix that computes the exclusive-or.

As a reference to facilitate the calculations, let us establish these two important properties of the exclusive-or:

$$X(u \otimes s) = Nu, \; X(u \otimes n) = u,$$

remembering that $X(u \otimes v) = X(v \otimes u)$.

7.2 Derivatives of monadic operations

Its calculation is immediate from definition (7.2)

i) $\quad \dfrac{\partial Iu}{\partial u} = X(Is \otimes In) = s,$

ii) $\quad \dfrac{\partial Nu}{\partial u} = s,$

iii) $\quad \dfrac{\partial Ku}{\partial u} = n,$

iv) $\quad \dfrac{\partial Ku}{\partial u} = n.$

We see that (i) and (ii) show that the transition from the variable s to n produces a change in the result. The property (ii) also indicates that the logical negation is not assimilable to the subtraction sign of the real variables, since (ii) it is not the negation of (i). The derivatives (iii) and (iv) have similarities to the derivative of a constant in the calculation of real variable functions, where $d[\text{Constant}]/dx = 0$.

7.3 Derivatives of dyadic operations

Here we have two partial derivatives for each logical function. Suppose, $u, v \in \Pi$ and let us look at these derivatives for the basic connectives.

Conjunction

$$\frac{\partial C(u \otimes v)}{\partial u} = X[C(s \otimes v) \otimes C(n \otimes v)] = X(v \otimes n) = v,$$

$$\frac{\partial C(u \otimes v)}{\partial v} = u.$$

Disjunction

$$\frac{\partial D(u \otimes v)}{\partial u} = Nv; \qquad \frac{\partial D(u \otimes v)}{\partial v} = Nu.$$

Implication

$$\frac{\partial L(u\otimes v)}{\partial u} = Nv; \qquad \frac{\partial L(u\otimes v)}{\partial v} = u.$$

Sheffer's connective

$$\frac{\partial S(u\otimes v)}{\partial u} = v; \qquad \frac{\partial S(u\otimes v)}{\partial v} = u.$$

Peirce's connective

$$\frac{\partial P(u\otimes v)}{\partial u} = Nv; \qquad \frac{\partial P(u\otimes v)}{\partial v} = Nu.$$

Equivalence

$$\frac{\partial E(u\otimes v)}{\partial u} = X(v\otimes Nv); \qquad \frac{\partial E(u\otimes v)}{\partial v} = X(u\otimes Nu).$$

Or-exclusive

$$\frac{\partial X(u\otimes v)}{\partial u} = X(v\otimes Nv); \qquad \frac{\partial X(u\otimes v)}{\partial v} = X(u\otimes Nu).$$

It is interesting to note that because $X(I\otimes N) = E$, the derivatives of E and X, share the following structure: $E(w\otimes w)$, for $w = u,v$. This implies that these derivatives equal s if $w \in \{s,n\}$.

From this catalog of derivatives of basic functions, both monadic and dyadic, some conclusions can be drawn. Operations that are negation of others ($N = NI$, $S = NC$, $P = ND$, $X = NE$) have the same derivative. This is easily demonstrable in general, so we can establish the following Lemma:

Lemma 7.1

$$\frac{\partial Op(u)}{\partial u} = \frac{\partial NOp(u)}{\partial u}.$$

The proof is immediate, given equality $X = X(N\otimes N)$.

Another fact to be highlighted, as already mentioned above, is the disappearance of the variable from which it is derived.

In the differential calculus of real variable functions, the derivative acts as a linear operator since it meets the following two conditions:

a) $\quad \dfrac{d}{dx}[kG(x)] = k\dfrac{d}{dx}G(x),$

b) $\quad \dfrac{d}{dx}[G(x) + H(x)] = \dfrac{d}{dx}G(x) + \dfrac{d}{dx}H(x).$

An interesting problem is whether the logical derivative (using logical versions of multiplication by a constant and addition) is a linear operator.[1] To focus on this, we will assume: (1) that the variables, $v \in \Pi$; (2) that the constant $t \in \{s, n\}$; (3) that multiplication is represented by the conjunction C; and (4) that addition is represented by the exclusive disjunction X. Under these conditions, given operations $Op(u)$ y $Op'(u)$, a (restricted) version of linearity could be defined by the following equalities:

a') $\quad \dfrac{\partial}{\partial u} C[t \otimes Op(u)] = C\left[t \otimes \dfrac{\partial}{\partial u} Op(u)\right]$

b') $\quad \dfrac{\partial}{\partial u} X\left[Op(u) \otimes Op'(u)\right] = X\left[\dfrac{\partial}{\partial u} Op(u) \otimes \dfrac{\partial}{\partial u} Op'(u)\right].$

To prove a'), you can do successively $t = s$ and $t = n$, and then check that, for each of these cases, the two members of the equation are equal. The proof of b') is more complex as it requires proving the following equality:

$$X[X(a \otimes b) \otimes X(c \otimes d)] = X[X(a \otimes c) \otimes X(b \otimes d)] \tag{7.3}$$

with $a = Op(s)$, $b = Op'(s)$, and $c = Op(n)$, $d = Op'(n)$. We have established that the $Op(u)$ functions y $Op'(u)$ are of the general form given by Eq. (7.1) and that the vectors belong to the set Π. Therefore, the outputs of the operations also belong to Π. Let us now define; $a = \alpha s + (1 - \alpha)n$; $b = \beta s + (1 - \beta)n$; $c = \gamma s + (1 - \gamma)n$; $d = \delta s + (1 - \delta)n$. Let us represent the output of each X as a probabilized vector, such as $\varphi s + (1 - \varphi)n$. Then the scalar projection of the output associated with matrix eq. (7.3) is

$$\varphi[\varphi(\alpha, \beta), \varphi(\gamma, \delta)] = \varphi[\varphi(\alpha, \gamma), \varphi(\beta, \delta)]. \tag{7.4}$$

Equation (7.4) has the general form of a functional equation whose unknown is the function φ and which is known as the bisymmetric equation (Aczél 1966). In our case, we know what the structure of the function is φ because it is the scalar projection of the operation of the matrix exclusive-or X. This general form

$$\varphi(x, y) = x + y - 2xy$$

satisfies the bisymmetric equation, which can be proved directly by substituting the in φ eq. (7.4).

7.4 Cross-derivatives

Given an operation $Op(u, v)$, $u, v \in \Pi$, it is true that

$$\frac{\partial}{\partial v}\left(\frac{Op(u, v)}{\partial u}\right) = \frac{\partial}{\partial u}\left(\frac{Op(u, v)}{\partial v}\right). \tag{7.5}$$

The general demonstration (Mizraji 2015) rediscovers the vector version of the bisymmetric equation (7.3) and the procedure is analogous to the one shown above. Below we will show the result of the cross derivatives for the fundamental dyadic operations. We will use the following notation to designate both cross-derivatives

$$\frac{\partial^2 \, Op(u, v)}{\partial[u, v]}.$$

It is easy to verify from the results shown above on the first derivaives that the cross derivatives of C, D, L, S, and P coincide:

$$\frac{\partial^2 C(u \otimes v)}{\partial[u, v]} = \frac{\partial^2 D(u \otimes v)}{\partial[u, v]} = \frac{\partial^2 L(u \otimes v)}{\partial[u, v]} = \frac{\partial^2 S(u \otimes v)}{\partial[u, v]} = \frac{\partial^2 P(u \otimes v)}{\partial[u, v]} = s.$$

The cross-derivatives of E and X also coincide:

$$\frac{\partial^2 E(u \otimes v)}{\partial[u, v]} = \frac{\partial^2 X(u \otimes v)}{\partial[u, v]} = n.$$

7.5 Successive derivatives

Successive derivatives do not seem to make sense in the Boolean derivation since the variable disappears in the first derivation. But if the variables belong to the set Π, an unexpected property arises that we illustrate below.

Let us first establish the following Lemma

Lemma 7.2

For $z, u \in \Pi$, with z independent of u, there is a derivative of z with respect to u, which is given by

$$\frac{\partial z}{\partial u} = X(z \otimes z).$$

Proof. For all pairs $z, u \in \Pi$, we have

(a) $Ku = s$,
(b) $C(s \otimes z) = z$,

In view of these results, we can assume the following equality:

$$\frac{\partial z}{\partial u} = \frac{\partial C(Ku \otimes z)}{\partial u} = X[C(Ks \otimes z) \otimes C(Kn \otimes z)] = X(z \otimes z).$$

This concludes the proof. Note if z is equal to the Boolean vectors s or n, then the derivative is equal to n, as is expected from the derivative of a constant. But if the vector is probabilistic, $z = \omega s + (1 - \omega)n$, the expected result for the scalar projection onto the vector s is:

$$s^T \frac{\partial z}{\partial u} = s^T X(z \otimes z) = 2\omega - 2\omega^2 = 2\omega(1 - \omega).$$

Let us use this Lemma 7.2 to analyze the successive derivative for a logical operation that depends on three variables $Op_1(u, v, w)$ with $u = \alpha s + (1 - \alpha)n$; $v = \beta s + (1 - \beta)n$; $w = \gamma s + (1 - \gamma)n$. Under these conditions, it is as follows:

$$\frac{\partial Op_1(u, v, w)}{\partial u} = Op_2(v, w) = f(\beta, \gamma)s + [1 - f(\beta, \gamma)]n.$$

The operation $Op_1(u, v, w)$ represents any logical operation of three variables (such as $D[E(u \otimes v) \otimes w]$ or $L[L(u \otimes v) \otimes E(Nu \otimes w)]$). Let us define the second derivative with respect to u as follows:

$$\frac{\partial}{\partial u}\left(\frac{\partial Op_1(u, v, w)}{\partial u}\right) = \frac{\partial^2 Op_1(u, v, w)}{\partial u^2} =$$

$$X[Op_2(v, w) \otimes Op_2(v, w)] = f'(\beta, \gamma)s + [1 - f'(\beta, \gamma)]n,$$

being $f'(\beta, \gamma) = 2f(\beta, \gamma)[1 - f(\beta, \gamma)]$.

It can be seen that this implies that if there is uncertainty represented by the probabilistic weights of the remaining vectors, the successive derivatives with respect to u iterate the process shown, and the evolution can be evaluated by the dynamics of the following equation to differences: $\varepsilon_{n+1} = 2\varepsilon_n(1 - \varepsilon_n)$. You can see that if $n \to \infty$, $\varepsilon_n \to 1/2$ which is the situation of maximum uncertainty.

7.6 Derived from basic tautologies

We will analyze here the relationships that differentiation establishes between the excluded middle, the *modus ponens* and the hypothetical syllogism.

1) *Excluded middle*: $p \vee \neg p$ (or $\neg p \vee p$)
Let us call the matrix version EM (for "excluded middle") and write down:

$$EM(u) = D(u \otimes N u).$$

Its derivative is

$$\frac{\partial EM(u)}{\partial u} = Ns.$$

2) *Modus ponens:* $[p \wedge (p \Rightarrow q)] \Rightarrow q.$
   Its matrix version is

$$MP(u, v) = L\{C[u \otimes L(u \otimes v)] \otimes v\},$$

and its partial derivatives are

$$\frac{\partial MP(u, v)}{\partial u} = NL(v \otimes v) = ND(Nv \otimes v) = N[EM(v)],$$

$$\frac{\partial MP(u, v)}{\partial v} = NL(u \otimes u) = N[EM(u)].$$

3) *Hypothetical syllogism:* $[(p \Rightarrow q) \wedge (q \Rightarrow r)] \Rightarrow (p \Rightarrow r).$
   We call its matrix version HS:

$$HS(u, v, w) = L\{C[L(u \otimes v) \otimes L(v \otimes w)] \otimes L(u \otimes w)\}$$

Here the derivatives with respect to u and w are symmetrical (for details, see Mizraji 2015):

$$\frac{\partial HS(u, v, w)}{\partial u} = N[MP(v, w)],$$

$$\frac{\partial HS(u, v, w)}{\partial w} = N[MP(u, v)].$$

The derivation with respect to the central variable has the following properties:

$$\frac{\partial HS(u, v, w)}{\partial v} = F(u, w).$$

$$F(u, s) = n = Ns; \; F(u, n) = N[EM(u)],$$

$$F(s, w) = N[EM(w)]; \; F(n, w) = n = Ns.$$

The Boolean derivative of a tautology is necessarily 0, whatever the form of derivation. However, in this matrix format, these Boolean "zeros" have a structure that reveals an interesting hierarchy among the tautologies we have analyzed. If we pre-multiply the partial derivatives by the operator N, the following diagram can represent the hierarchy shown by the derivatives:

$$HS(u, v, w) \xrightarrow{N\partial/\partial u} MP(v, w) \xrightarrow{N\partial/\partial v} EM(w) \xrightarrow{N\partial/\partial w} s$$

Note that the derivation with respect to the central variable of HS produces a jump, not included in the diagram, which bypasses MP and connects with EM or s.

Before we end this section, let us discuss that there is another way to define the derivative for logical operations with variables belonging to the set Π. This definition is as follows:

$$\frac{\delta Op(u)}{\delta u} = X[Op(u) \otimes Op(Nu)].$$

This derivative is appropriate to analyze the effect of changes for vector variables with probabilistic weightings in the open interval $(0,1)$ and it is a derivative that does not make the variable with respect to which it is derived disappear. It is a topic with potentially interesting results, but we will not explore it here.

7.7 Logical integrals

The integrals that we will deal with here are the antiderivatives of logical operations.

7.7.1 Definition: Boolean integral

Given a vector logic function Op, we define the Boolean integral as another logical function Υ such that

$$\frac{\partial \Upsilon}{\partial t} = Op \tag{7.6}$$

where $t \in \Pi$ is a logical variable not included in Op. We will define two types of integrals that satisfy eq. (7.6). We will call the first one "general integral" and write it down as

$$\Upsilon = \int Op \partial t . \tag{7.7}$$

The symbol ∂t represents the new variable with respect to which it is integrated. As can be seen, this integral extends the number of variables of the operand, the opposite of what happens with differentiation that eliminates the variable from which it is derived.

The following theorem shows us that there is a well-defined procedure for constructing a general integral.

Theorem 7.1.
An Op logic operation supports a general integral of the form

$$\Upsilon = \int Op \partial t = HL\left(Op \otimes H't\right); \qquad H, H' \in \{I, N\}. \tag{7.8}$$

Proof. Note that $L(Op \otimes s) = s$ and $L(Op \otimes n) = NOp$. The derivative of the general integral with respect to t is

$$\frac{\partial \Upsilon}{\partial t} = X\left[HL(Op \otimes H's) \otimes HL(Op \otimes H'n)\right] =$$

$$X(H \otimes H)\left[L(Op \otimes H's) \otimes L(Op \otimes H'n)\right].$$

As $X(I \otimes I) = X(N \otimes N) = X$, for any combination of $H, H' \in \{I, N\}$ finally results in

$$\frac{\partial \Upsilon}{\partial t} = X[L(Op \otimes s) \otimes L(Op \otimes n)] = X(s \otimes NOp) = Op.$$

This proof shows that the general integral can take various forms; here are some of them:

$\{1\}\ \displaystyle\int Op\ \partial t = L(Op \otimes t),$

$\{2\}\ \displaystyle\int Op\ \partial t = NL(Op \otimes t),$

$\{3\}\ \displaystyle\int Op\ \partial t = NL(Op \otimes Nt) = C(Op \otimes t),$

$\{4\}\ \displaystyle\int Op\ \partial t = L(Op \otimes Nt) = NC(Op \otimes t).$

A consequence of the law of contraposition of implication is the following corollary:

$$\int NOp\ \partial t = L(NOp \otimes Nt) = L(t \otimes Op).$$

Note that this general integral is valid regardless of the structure of the Op.

Now we will see that there is another integral that satisfies definition (7.6) but that, unlike the general integral, depends on the structure of the operand Op. We will call it a "particular integral" and we will symbolize it as

$$\Upsilon_P = \int_P Op \partial t. \tag{7.9}$$

Suppose that the *Op* operation can be expressed as follows:

$$Op = Op(a[1], a[2], \ldots, a[n]).$$

In this expression, the indices i indicate the position in the formula, so two variables with different indices can coincide For example, in an expression such as

$$L[C(u \otimes v) \otimes E(u \otimes v)]$$

is $a[1] = u$, $a[2] = v$, $a[3] = u$, and $a[4] = v$.

The procedure for constructing a particular integral consists of substituting the $a[i]$ by logical functions containing the variable with respect to which it is integrated:

$$a[i] \rightarrow B_i(a[i], t),$$

where $B_i(a[i], t)$ be such that when derived from t, return Op to the original format. We will illustrate this for the particular case of two-variable Op functions, where the conditions that must be met $B_i(a[i], t)$ can be easily established.

Given a two-variable vector logic operation $Op(u, v)$. Let us establish by means of the following Lemma a sufficient condition for the functions of the $B(u, t)$ and $B'(v, t)$ generate particular integrals.

Lemma 7.3.
Substitutions $u \rightarrow B(u, t)$ and $v \rightarrow B'(v, t)$ within the logical function $Op(u, v)$ generate a particular integral when one of the following properties is met:

$$a)\ B(u, s) = u;\quad B'(v, s) = v;\quad Op\big[B(u, n), B'(v, n)\big] = n,$$

$$b)\ B(u, n) = u;\quad B'(v, n) = v;\quad Op\big[B(u, s), B'(v, s)\big] = n.$$

Proof. The evaluation of the Boolean derivatives in situations (a) and (b) generate the conditions $X[Op(u, v) \otimes n] = X[n \otimes Op(u, v)] = Op(u, v)$.

We illustrate this with the following examples, where we show the general and particular integrals.

Example 1. $Op(u, v) = D(u \otimes Nu)$.

General integral: $\Upsilon = \displaystyle\int D(u \otimes Nu)\partial t = L[D(u \otimes Nu) \otimes t]$,
Particular integral: $u \xrightarrow{} C(u \otimes t);\ Nu \rightarrow C(Nu \otimes t)$,

$$\Upsilon_P = \int_P D(u \otimes Nu)\ \partial t = D[C(u \otimes t) \otimes C(Nu \otimes t)].$$

It is easy to see that these substitutions of the particular integral satisfy item (a) of Lemma 7.3.

Example 2. $Op(u, v) = L(u \otimes Nv)$.

General integral: $\Upsilon = \displaystyle\int L(u \otimes Nv)\partial t = L[L(u \otimes Nv) \otimes t]$,
Particular integral: $u \xrightarrow{} D(u \otimes t);\ v \rightarrow D(v \otimes t)$,

$$\Upsilon_P = \int_P L(u \otimes Nv)\ \partial t = L[D(u \otimes t) \otimes ND(v \otimes t)].$$

In this case, we are in the situation of item (b) of Lemma 7.3.

Similar to what is shown by the differential and integral calculus of continuous variables, this discrete calculus also exhibits a degradation of complexity when deriving and an increase in complexity when "injecting" information during integration. We would like to conclude by pointing out a topic that deserves further reflection. It is the analogy of the relations between information and complexity in differential and integral calculus with the relations between energy and complexity in thermodynamics. As thermodynamics teaches us, spontaneous processes tend to reduce the complexity of systems until equilibrium is reached, and increasing complexity requires an injection of energy (or low-entropy energy, as some physicists emphasize). In this analogy, thermodynamic spontaneity is metaphorically similar to the completely algorithmic character of derivation, and the injection of energy, correlative to the increase in physical complexity, is also metaphorically similar to the need to incorporate information to achieve the increase in mathematical complexity that occurs in integration. At a time when so much research is being done on the thermodynamics of information, this analogy, intuited by so many researchers, deserves to be highlighted.

It is worth mentioning Lukasiewicz's important observations on "generalizing deduction" (Lukasiewicz, 1931). He describes how deduction can increase the generality or complexity of formal expressions and demonstrates that, in certain cases, it is possible to move deductively from the particular to the general. This suggestive discovery by Lukasiewicz invites us to consider possible links between methods of constructing particular integrals for logical functions and the idea of generalizing deduction.

References

Aczél, J. (1966) Lectures on Functional Equations and their Applications, Academic Press, New York.

Couturat, L. (1914) L'Algèbre de la Logique (Collection Scientia, Phys-metématique No 24) (Second Edition), Gauthier-Villar, Paris. 1914 [Reedition; Albert Blanchard, Paris, 1980].

Davio, M., Deschamps, J.P., Thayse, A. (1978) Discrete and Switching Functions, McGraw Hill, New York.

Gorbatov, V.A. (1988) Fundamentos de la Matemática Discreta, Mir, Moscow.

Lukasiewicz, J. (1931) Comments on Nicod's axiom and on "generalizing deduction" published, in Łukasiewicz, J., Selected Works 1970, (L. Borkowski, ed), North Holland, Amsterdam.

Melzak, Z.A. (1983) Bypasses: A Simple Approach to Complexity, Wiley, New York.

Mizraji, E. (2013) En Busca de Las Leyes Del Pensamiento (Second Edition), Trilce-Dirac, Montevideo.

Mizraji, E. (2015) Differential and integral calculus for logical operations: A matrix-vector approach, Journal of Logic and Computation, 25: 613–638.

Shannon, C.E. (1940) A Symbolic Analysis of Relay and Switching Circuits, MSc Thesis, MIT.

Shannon, C.E. (1938) A Symbolic Analysis of Relay and Switching Circuits, Transactions American Institute of Electrical Engineers, 57: 38–80. (abbreviated version of the Master's Thesis).

Tucker, J.H., Tapia, M.A., Bennett, A.W. (1998) Boolean integral calculus, Applied Mathematics and Computation, 26: 201–236.

Vichniac, G.Y. (1990) Boolean derivatives on cellular automata, Physica D, 45: 63–74.

Whitehead, A.N. (1898) A Treatise on Universal Algebra, Cambridge University Press, Cambridge. [2009 Reissue].

Chapter 8
Binary cellular automata and the complexity of the simple

When Martin Gardner published in October 1970, in his "Mathematical Games" section of *Scientific American*, his description of the so-called Game of Life (usually "Life"), the extent of the consequences that Life would bring was perhaps not foreseeable. A tumult of readers began, armed with graph paper, pencil and eraser, to explore the fate of Life's configurations. There was nothing random about Life because its creator had established three simple laws that stipulated the conditions of survival, death, and birth for the cells of the grid space in which the game took place, but despite these well-defined laws, the fate of the configurations could be as unpredictable as if it were a random process. This fact began to upset a philosophical conviction that lay at the back of the minds of many scientists. Guided by this conviction, many believed that the ultimate goal of science was to find the fundamental laws that governed our cosmos, with the certainty that, once those laws were possessed, the secrets of the universe would be revealed. A disturbing idea began to take hold from the experience of exploring the evolution of Life's configurations: this idea indicated that even if we knew the ultimate laws precisely (in a similar way to what happened in Life), our ability to understand and predict would continue to be severely limited. This was not an entirely new idea. In fact, in the late nineteenth and early twentieth centuries, mathematicians and astronomers such as Henri Poincaré and Karl Sundman who were investigating the three-body problem clearly understood that the laws of gravitational interaction between three bodies led to situations where prediction of dynamics was difficult or impossible (Poincaré 1890; Sundman 1912). Now, after Life, the massive experience with a system whose laws did not require the sophisticated mathematics of the three-body problem, made many scientists understand those limitations that had previously been understood only by a few experts.

Life was created by mathematician John Horton Conway, known first and foremost as a virtuoso practitioner of pure mathematics. Life takes place in a gridded space that can be limitless, or extensive and with the torus topology (where the upper edge connects to the lower edge and the right edge continues to the left), or merely with enough cells to give room to experiment away from the edges. In that gridded space, each "cell" has a neighborhood of eight cells and it is from that neighborhood that everything depends, according to the laws of Life. This entity that runs in a space of cells was called a "cellular automaton." A stuffed cell is "alive." Several living cells generate a configuration. These configurations evolve in a discrete time, T, $T+1$, $T+2$, etc., with $T=0$ being the initial instant. The three laws of the universe where Life unfolds are the following (Gardner 1970, 1985):

https://doi.org/10.1515/9783112230053-009

1) *Survival*: Every living cell that has two or three living cells adjacent to it survives and is passed on to the next generation.

2) *Death*. Every living cell that has four or more neighbors dies from overcrowding. Living cells that are isolated or have only one living cell in their environment die from isolation.

3) *Birth*. Each empty cell, adjacent to exactly three living cells, becomes alive in the next generation.

The dimension of the impact of this game from 1970 onward can be assessed by a comment by Martin Gardner: "My 1970 column on Conway's 'Life' met with such an instant enthusiastic response among computer hackers around the world that their mania for exploring 'Life' forms was estimated to have cost the nation millions of dollars in illicit computer time" (Gardner 1983). This phrase, which Gardner included in a book, comes from a time when the use of computer time was extremely expensive. In those days before our today personal computers, large universities had a powerful central computer and the laboratories that used it had to have a considerable budget to pay for calculation time. But the important thing about Gardner's comment is that it implicitly indicates the large volume of research that went into exploring the properties of Life. Some institutions, such as MIT, formalized these explorations and their researchers found configurations with extraordinary behaviors; some of them are described in the chapters of Gardner's book (1983) and are usually incorporated into the menus of programs found on the web.

One of the most illustrious configurations, known for its unpredictable behavior, is called R-pentomino, a small structure with five living cells. Its evolution is one of the surprises that Life has in store for us. In Figures 8.1 and 8.2, we show this configuration at the initial instant and what happens some time later.

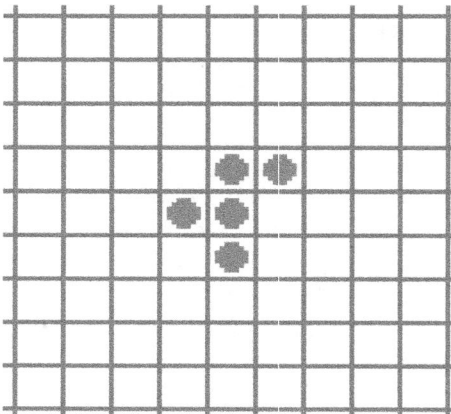

Figure 8.1: R-pentomino at $T = 0$; population = 5. This surprising configuration, consisting of only five living cells, displays an evolutionary richness that extends over hundreds of generations, producing a wide variety of new configurations.

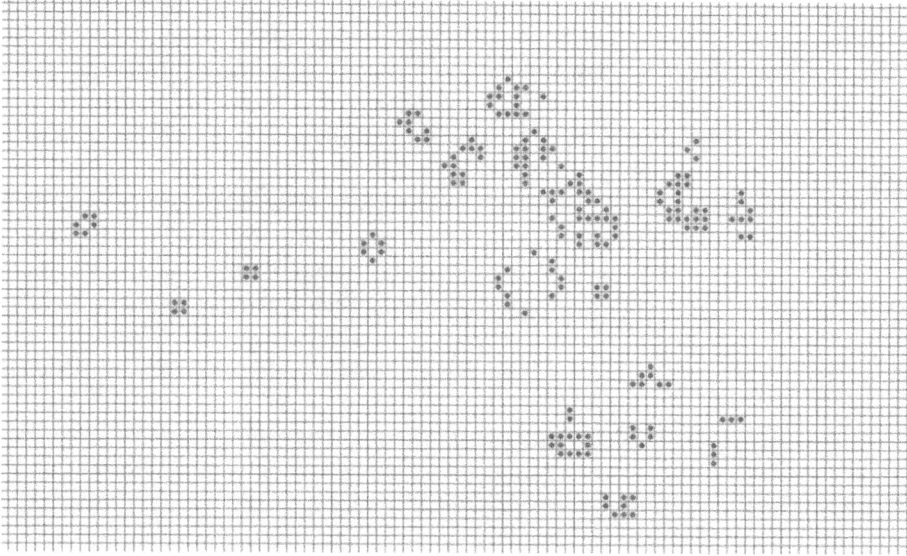

Figure 8.2: R-pentomino at $T = 572$; population = 210. Here we show that the 5 living cells in the initial configuration, as shown in Figure 8.1, have evolved by generation 572 into a population of 210 living cells. It should be noted that the Winlife 32 program has a limited size and absorbent edges, and that more powerful software can display greater richness. [Figures 8.1 and 8.2 obtained with Winlife32, an open-access program.] Some configurations counted by the program were not included in the grid shown in the figure.

As can be understood by looking at these figures, the small initial configuration of five living cells triggers a very complex evolution. Is it possible to predict, knowing Conway's laws, that R-pentomino will reach that state of 210 living cells and the details of its multiple configurations for $T = 572$? So far no one has succeeded.

To get an idea of the variety of configurations that Life supports, let us imagine a cell space of N cells. If we are dealing with a gridded space, of 100×100 then $N = 10{,}000$. What is the variety of configurations in that space? If we count all the cases with only one live cell, the cases with only two, and so on up to N, we have

$$\binom{N}{1} + \binom{N}{2} + \cdots + \binom{N}{N} = 2^N - 1,$$

where

$$\binom{N}{i} = \frac{N!}{i! \ (N-i)!}$$

is the combinatorial number that tells us how many sets of i living cells can be formed from N cells. For $N = 10,000$, the variety of possible configurations is on the order of 10 to the exponent 3,000. There are a few configurations whose fate is foreseeable. If we go to the extreme of imagining a single living cell in T or a set of isolated cells, then in $T + 1$, none will be left alive; if we go to the other extreme and assume N live cells in T, in $T + 1$, there will be no living cells either.

Life's connection with cosmology was not long in coming. Life's gridded space is a discrete universe, with three well-defined laws and configurations that can leave a plethora of varied objects during its evolutionary trajectory, some stable and others pulsating (such as three aligned and adjacent living cells that pulse between the horizontal and vertical), and mobile elements (such as sliders, or "gliders"). This is a universe with a Supreme Creator, Conway, and with little creators who are us, the ones who invent configurations at $T = 0$.[1]

A book written by William Poundstone, *The Recursive Universe* (1987), shows an interesting counterpoint between Life events obtained through computational experiments described in the even-numbered chapters, and elements of physical cosmology presented in the odd-numbered chapters.

But the cosmological approach went further, and led to a deep scientific problem: how to understand the origin of the extreme complexity shown by the dynamics of Life? In the last lines of the last chapter of his 1983 book, Martin Gardner writes: "Now we are back to Leibniz and his stupendous vision of a transcendent Mind, contemplating all possible games, then choosing for our universe the Game that best suits the Mind's incomprehensible purposes." The reader interested in the origin of this reflection can read the ways in which Leibniz expressed these ideas (texts where Gardner's Mind is Leibniz's God) in items 53, 54, and 55 of his Monadology (Leibniz 1714), and in a more detailed way in his Theodicy (Leibniz 1710), where Leibniz amuses himself (along with the reader) with theological polemics that evoke a story by Jorge Luis Borges. The emphasis we are placing on this quotation concerning Leibniz is because he implicitly proposed a way of solving the problem of the origin of the complexity of a cellular automaton like Life. In the Monadology is the idea of a Mind capable of fully exploring the set of universes. Under this view, the three laws of Life define one of the many possible universes in which the laws depend on the cell and its immedi-

---

1 As an interesting and curious point, it turns out that at the time of writing this (2025) one of the generative Artificial Intelligence systems summarizes these properties of Life in the following way: "Conway's Game of Life is a cellular automaton played on a two-dimensional grid of square cells. Updates occur in discrete time steps, with each cell's state determined by its neighbors. Neighboring cells include the eight adjacent ones – horizontal, vertical, and diagonal. Each cell follows three simple rules: survival, birth, and death. Given these rules, patterns can evolve into stable structures, oscillators, or moving entities. Uniform initial conditions may lead to symmetry, but randomness often yields surprising complexity. Nontrivial behaviors like gliders and guns demonstrate emergent computation. Despite its simplicity, the system is Turing complete, capable of simulating any algorithm. As a model, it illustrates how complexity can arise from local interactions and deterministic rules."

ate environment. But unlike the theological universes, here the number of universes in the Life family can be known. We will see what that number is. The first point to note is that in these universes the fate of each cell depends on its immediate environment and eventually on itself (the latter does not occur in Conway's universe laws. The second point is that each cell can be in only two states: empty or alive. A natural numerical representation for such an automaton is: 0 = empty, 1 = alive. Therefore we have the automaton representable by Boolean variables. On the other hand, the fate of each cell unfolds in a discrete time. Then the laws of each of these automata are condensed into a number of binary logical operations, which can always be represented by a table and also by a Boolean function of the form

$$y(T+1) = F(y(T), x_1(T), \ldots, x_8(T)) \tag{8.1}$$

where $y(T)$ is the Boolean variable that represents the cell we want to compute and we represented by $x_1(T), \ldots, x_8(T)$ the variables of the environment. In this context, referring to the family of automata such as Conway's, the maximum of cellular universes with laws of immediate adjacency is given by the general expression shown in Chapter 6. Let us call $C(9)$ that number that gives us the variety of Boolean functions of nine variables as in eq. (8.1). The result is

$$C(9) = 2^{(2^9)} \cong 1,34 \times 10^{154}.$$

Conway's creative mind managed to pick Life out of that tumult of universes and saw that it was good for humans. But it possibly true even now, 2025, that there does not seem to be a human mind that can see, and eventually explore with its computer, each of these 10^{154} universes. At the same time, this scale seemed to put the investigation of the origin of complexity in cellular automata at an *impasse*, until the talent of Stephen Wolfram entered this scenario.

8.1 Elementary cellular automata

Wolfram is a physicist born in England in 1959, who at the age of 15 began to investigate topics of nuclear physics, a subject in which he had his first publications in 1975 and 1976. He continued to publish papers on particle physics and in late 1979 earned his Ph.D. from the California Institute of Technology before a jury of which Richard Feynman was a member. But after this promising start in the field of particle physics, Wolfram's intelligence is captured by the theoretical and computational potentialities of the world of cellular automata. In 1983 he published in Reviews of Modern Physics an extensive article entitled "Statistical mechanics of cellular automata." This pioneering article and other relevant articles, as well as appendices with much supplementary information, are compiled in the book "Cellular Automata and Complexity" (Wolfram 1994). This book also includes as an appendix the complete list of Wolfram publications up to 1991. After 1986, Wolfram concentrated on developing his *Mathematica* program and developing his company. In 2002 he published an extensive and debated book called *A New Kind of Science* (Wolfram 2002). One of the basic ideas present in that book is that the basic laws of science in the immediate future will not be the traditional equations of physics (which will nevertheless maintain their validity and power at their own level of observation) but small logical programs such as those that govern cellular automata.

Wolfram's great creation occupies a large part of his 1983 article on the statistical mechanics of cellular automata. This creation was baffling and amazing. It consisted of the invention of a family of binary automata, which instead of "living" in a two-dimensional cellular space, where each cell had an environment of eight neighbors, lived in a one-dimensional cellular space where each cell had only two immediate neighbors. Each of these one-dimensional automata corresponded to a Boolean function with the structure

$$x_0(T+1) = F_i(x_{-1}(T), x_0(T), x_1(T)). \tag{8.2}$$

where F_i is one of the possible laws governing that set of automata universes; x_{-1} and x_1, respectively, are the states of the left and right neighbors of x_0. These laws are the Boolean functions of three variables, such as those we studied in the previous chapters, and correspond to a three-variable truth table or a logical formula of structure

$$y' = f(x, y, z), \qquad x, y, z \in \{0, 1\}.$$

How many such functions are there? By applying our well-known formula, we have

$$C(3) = 2^{(2^3)} = 256.$$

If there is no human being who, like Leibniz's God, can extract laws and systematize them and choose from all the Boolean 10^{154} universes of the Conway class, Wolfram instead created a set of universes that can be exhaustively investigated. But surprise and amazement arise when it is found that in this set of 256 universes, there are several in which the temporal evolution of the configurations is as unpredictable as the evolution of the configurations of Life.

The laws that govern each of these 256 automata can be represented in various ways. Representation as truth tables has been important. From them, Wolfram devised a compact method to fully characterize each of the individual automaton Here is the general format of those tables, where the symbols (*a b c d e f g h*) are Boolean variables. This is the general format:

111	110	101	100	011	010	001	000
a	*b*	*c*	*d*	*e*	*f*	*g*	*h*

Note that the given automaton governed by this table is also representable by a Boolean polynomial $\mathrm{Aut}(x, y, z)$:

$$\text{Aut}(x,y,z) = axyz + bxy(1-z) + cx(1-y)z + dx(1-y)(1-z) +$$
$$e(1-x)yz + f(1-x)y(1-z) + g(1-x)(1-y)z + h(1-x)(1-y)(1-z) \tag{8.3}$$

with $x,y,z \in \{0,1\}$. In Wolfram nomenclature, each automaton is assigned the decimal number associated with the binary number produced by the outputs of the table: *abcdefgh*. Thus, the number of the automaton $W(\text{dec})$ is given by the decimal

$$\text{dec} = a2^7 + b2^6 + c2^5 + d2^4 + e2^3 + f2^2 + g2^1 + h2^0.$$

Rule 30, or $W(30)$, corresponds to the 00011110 outputs. This shows that we are dealing with an ingenious code, where the decimal that characterizes a rule, once transformed into a binary, makes explicit the law of the automaton. The rule of the automaton $W(110)$ corresponds to the binary number 01101110, which, when taken to the table gives us the law of transformation. These automata are now called elementary cellular automata (ECA). A more technical name is automata of class $k=1, r=2$, where r is the number of states a cell can display and k is the radius of influence of the environment. ECA is binary and its environment of influence is only each of its neighbors. With the increase in computer power, more complex automata have been investigated (e.g.) $k=2, r=2$, which increases the number of laws. Here we will only deal with ECA.

Because of the isotropy of cell space, which in this case represents right-left symmetry, and dynamic equivalence in the face of color change, Wolfram's set of 256 rules contains sets of automata of identical dynamics, except in pictorial details. In the ECA literature, transformations that generate automata of equivalent dynamics are called conjugation (*c*), reflection (*r*), and conjugation and reflection (*c–r*). Using the *abcdefgh* outputs of the Wolfram table, and representing the negation $\neg p = p'$. These transformations are defined as follows:

c) $h'g'f'e'd'c'b'a'$,
r) $aecgbfdh$,
r–c) $h'd'f'b'g'c'e'a'$.

The conjugation transforms all the cells of the set of ECA from u to u', causing negative images of the original. Reflection produces equivalent automata by left-right isotropy. To exemplify the reflection, let us look at the case of implication: $(11) \rightarrow 1$, $(10) \rightarrow 0$, $(01) \rightarrow 1$, $(00) \rightarrow 1$. The binary number is $abcd = 1011$. The reflection reverses the order of the pairs leaving $(11) \rightarrow 1$, $(01) \rightarrow 1$, $(10) \rightarrow 0$ $(00) \rightarrow 1$, and the associated binary is $1101 = acbd$. A very relevant consequence of these operations is that the total of 256 different rules is reduced to only 88 dynamically distinct rules. Wolfram represented the evolution of ECA by means of diagrams where the time in the space of the graph runs from top to bottom, and each horizontal line is a row of binary cells corresponding to a time T. The top row of each diagram shows the initial conditions for $T=0$. In Figures 8.3 and 8.4, we show for random initial conditions, the evolution of automata with rules 30 and 110, perhaps the most important and investigated ECA for their rich computational properties.

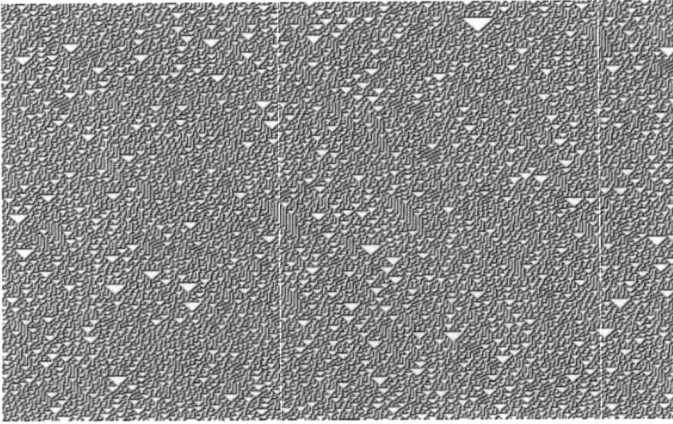

Figure 8.3: Evolution of automaton 30 from random initial conditions. The dynamic complexity of one of the most famous elementary cellular automata is shown here, where an initial line of (pseudo-)random states maintains this randomness homogeneous throughout the evolutionary history of the automaton. Here, time $T = 0$ is represented by the first row of cells, and downwards the history of that initial state develops in successive times $T = 1$, $T = 2$, etc.

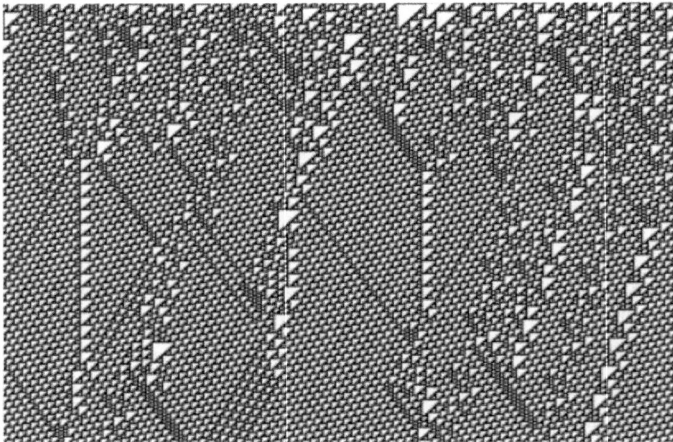

Figure 8.4: Evolution of the 110 automaton for random initial conditions. This elementary cellular automaton 110 is one of the most investigated structures for its computational potential. It contains a complex structure confined within an ordered framework, and this coexistence of order and disorder is the basis of the computational capabilities investigated by Wolfram's group.

Automaton 30 also has a very remarkable property, that is the complex way in which it evolves from an initial state in which only one cell is active in $T = 0$. We illustrate this in Figure 8.5.

Automaton 30 exhibits the surprising behavior shown in this figure, where a single live cell in the one-dimensional row generates a practically random display over successive times.

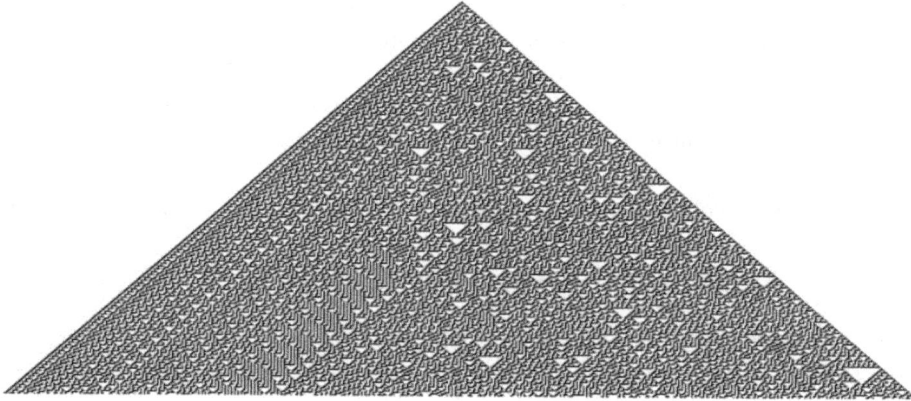

Figure 8.5: Evolution of automaton 30 from just one cell. Automaton 30 exhibits the surprising behavior shown in this figure, where a single live cell in the one-dimensional row generates a practically random display over successive times. It is very curious that the succession of black and white (or live or dead) cells in the vertical of the initial state is associated with a succession of 1 s and 0 s that satisfy very demanding pseudo-randomness tests. [Figures 8.3 to 8.5 were obtained using the program found at http://devinacker.github.io/celldemo.].

One-dimensional space can be considered infinite, annular, or finite and extensive, in which case (as in Life) there is an edge effect where the laws are invalidated.

Note that the rules of ECA, being third-degree Boolean polynomials, correspond to matrices representable by second-degree Kronecker polynomials. In the case of rules 30 and 110, the three versions of these polynomials (right, center, and left) are:

Rule 30: $A(30) = [NNNI) = (MNKI) = (MNKI]$
Rule 110: $A(110) = [NKKM) = (NIKI) = (NIKI]$

Another peculiar feature of rule 30 is the simplicity of the logical formula that represents it, as opposed to the complexity of its dynamics. Indeed, it can be proven (Mizraji 1996) that rule 30 corresponds to the following matrix:

$$A(30) = X(I \otimes D),$$

which in classical logic terms is the operation $[u \neq (p \vee r)]$.

Although the laws of ECA are triadic and bivalent logical functions, the theoretical study and prediction of dynamics is in many cases very difficult or impossible. Given this fact, the detailed investigation of the dynamics associated with the 256 rules was empirical, and was achieved after the development of computers with high memory and speed. Nowadays, there are programs on the web that compute the laws of the ECAs. Computational research under various initial conditions led Stephen Wolfram to classify the dynamics of ECA into four classes:

Class 1: The evolution of ECA leads to homogeneous states (e.g., rules 0 and 4).

Class 2: Evolution leads to sets of simple or periodic structures (e.g., rules 8 and 24).

Class 3: The evolution of the automata produces chaotic structures (e.g., rules 30 and 45).

Class 4: Evolution leads to complex localized structures (e.g., rule 110).

Wolfram's enigmatic finding was the behavior of ECA of classes 3 and 4. The chaotic character of automaton 30 was studied with the basic techniques of nonlinear dynamical systems, and showed extreme sensitivity to small variations of the initial conditions and Lyapunov exponents associated with exponential divergence. Moreover, Wolfram found that the sequence of binary numbers generated by automaton 30 for a fixed position (such as the vertical cells defined by the initial cell in Figure 8.5) had a more robust level of randomness (or, if one wants to be rigorous, pseudo-randomness because it is a deterministic process) than that of the congruential techniques frequently used by pseudorandom number generators (Wolfram 1994). The automaton 110 has been intensively researched and has been found to have the ability to implement a universal computer, that is, the possibility of being able to simulate any computer program and any calculation. The complicated demonstration of this universality was published (Cook 2004; Wolfram 2002). Wolfram has extended its classification to other spaces and considers Life an automaton, just like the 110, of class 4. Wolfram has created for class 3 and 4 automata the notion of "computational irreducibility," which states that in the face of random initial conditions, there are no simple ways to predict the dynamics, and that it can only be known by computing it.

References

Cook, M. (2004) Universality in elementary cellular automata, Complex Systems, 15: 1–40.

Gardner, M. (1970) The fantastic combinations of John Conway's solitaire game "life", Sci. Am, October.

Gardner, M. (1983) Wheels, Life and Other Mathematical Amusements, Freeman, New York.

Leibniz, G.W. (1714) The Monadology (translated by Robert Latta 1898; online: http://home.datacomm.ch/kerguelen/monadology/printable.html)

Leibniz, G.W. (1710) Theodicy, Published 1985 by Open Court Publishing Company.

Mizraji, E. (1996) The operator of vector logic, Math. Log. Quart, 42: 27–40.

Poincaré, H. (1890) Sur le problème des trois corps et les équations de la dynamique, Acta Mathematica, 13: 1–270.

Poundstone, W. (1987) The Recursive Universe, Oxford University Press, Oxford.

Sundman, K.F. (1912) Mémoire sur le problème des trois corps, Acta Mathematica, 36: 105–179.

Wolfram, S. (1994) Cellular Automata and Complexity, Addison-Wesley, New York.

Wolfram, S. (2002) A New Kind of Science, Wolfram Media. Inc. Winnipeg.

Chapter 9
Analyzing the internal structure of complex ECA through logical matrices: an unfinished exploration

The question that motivated the results we will show in this chapter is this: even assuming that a system is computationally irreducible, can there be "signatures" in the mathematical structure of the automaton law, which indicate that this law can generate a complex dynamic? This question has not been fully answered, but a construction we developed years ago, the logical spectra (Mizraji 2004), can bring us closer to a partial answer. For the calculation of these spectra, the Kronecker polynomials shown in Chapter 4 were used. But the inspiration for its construction came from an important computational structure called the "Fredkin gate." We will make a brief review of the properties of this logic gate, we will show its representation through the matrix algebra formalism of logical operators and we will also see its natural link with some rules of the elementary cellular automata (ECA).

9.1 Fredkin's gate and the activation of its computational capabilities

The Fredkin gate (Fr) is a truth table in which three binary inputs correspond to three binary outputs. A relevant property of this gate is its reversibility. This means that $Fr(abc) = xyz$ and $Fr(xyz) = abc$. Operating iteratively results in $Fr[Fr(abc)] = abc$. The other property, which is the key to its importance, is its latent capacity to compute conjunction, disjunction, implication, and negation. Activating these latent logic operations requires selecting one of its three outputs and forcing formatting on its inputs. The Fredkin gate produced strong interest because the suggestion that reversible computers could be built that would drastically reduce their power consumption. This Fredkin gate was inspired by the essential reversibility of the laws of physics (Fredkin and Toffoli 1982), and provided a theoretical basis to investigate some aspects of the thermodynamics of computing (Feynman 1999).

A compact definition of the Fredkin gate is as follows:

$$Fr(1ab) = 1ab; Fr(0ab) = 0ba$$

with $a, b \in \{0,1\}$. This produces a truth table with three inputs and three outputs per row, and a total of eight rows. The representation of this operation by means of matrix formalism clearly shows how the activation of the logical capacities of the Fredkin gate occurs (Mizraji 1996). We will not detail the calculations, as they are somewhat extensive and the reader can make them using the elements presented in Chapter 4. The first step is to express the table of eight rows, each with three inputs

https://doi.org/10.1515/9783112230053-010

and three outputs, as a matrix associative memory where the input and output varia-bles are represented by Q-dimensional column vectors of the set $\tau = \{s, n\}$. The varia-bles within each triplet are linked using Kronecker products. This results in a square matrix $Q^3 \times Q^3$ that after several factorizations is expressed as

$$F = ss^T \otimes I^{[2]} + nn^T \otimes R, \tag{9.1}$$

where I is the logical identity matrix, and R is a matrix of exchange of positions, with the structure

$$R = s \otimes I \otimes s^T + n \otimes I \otimes n^T. \tag{9.2}$$

The following properties can be demonstrated:

a) $R^2 = I^{[2]}$;
b) if $u, v \in \tau$ then $R(u \otimes v) = v \otimes u$;
c) $F^2 = I^{[3]}$.

Kronecker's power $I^{[3]}$ ensures iteration invariance:

$$F^2(u \otimes v \otimes w) = Iu \otimes Iv \otimes Iw = u \otimes v \otimes w.$$

Note that in this formalism of operators, the vectors u, v, and w could belong to the set Π of probabilized vectors.

To analyze how this gate operates, the first step is to select one of the three out-puts. We will call the first channel 1, channel 2 the second, and channel 3 the third. Below are three matrices that serve as channel selectors. These selectors for channels 1, 2, and 3 are, respectively, the matrices

$$\Phi_1 = I \otimes (s+n)^T \otimes (s+n)^T$$

$$\Phi_2 = (s+n)^T \otimes I \otimes (s+n)^T$$

$$\Phi_3 = (s+n)^T \otimes (s+n)^T \otimes I.$$

The pre-multiplication of F by one of the selector matrices generates a logical function with one output and three inputs. This is a triadic function to which we assign a second-degree Kronecker polynomial. Let us express the results that arise from the application of the three filters as right polynomials:

$$\Phi_1 F = [I\ I\ I\ I], \quad \Phi_2 F = [K\ I\ N\ M], \quad \Phi_3 F = [K\ N\ I\ M].$$

Activation of computing power occurs when one or two variables are forced into the input, causing s or n to act as constants. Channel 1 does not produce interesting re-sults. On the contrary, basic logical operators are produced by the other two channels, as shown by the following equations (Mizraji 1996):

Channel 2

$$C(u \otimes v) = \Phi_2 F(u \otimes v \otimes n); \quad D(u \otimes v) = \Phi_2 F(u \otimes s \otimes v)$$
$$L(u \otimes v) = \Phi_2 F(u \otimes v \otimes s); \quad Nu = \Phi_2 F(u \otimes n \otimes s)$$

Channel 3

$$C(u \otimes v) = \Phi_3 F(u \otimes n \otimes v); \quad D(u \otimes v) = \Phi_3 F(u \otimes v \otimes s)$$
$$L(u \otimes v) = \Phi_3 F(u \otimes s \otimes v); \quad Nu = \Phi_3 F(u \otimes s \otimes n).$$

These are the admirable results obtained by Fredkin using Boolean operations and diagrams. What is important for this chapter is that the three-variable functions obtained by selectors 2 and 3, which are three-variable functions, are also rules of the ECA space. And these rules store, hidden, dyadic logical functions. Seen from the perspective of the ECA, what results from filtering the Fredkin gate through channels 2 and 3 are Wolfram rules 202 and 172. It should be noted that both rules generate class 2 automata and, together with rules 216 and 228, they make up the same symmetry group. This way of computing by activating implicit logical operations suggested us to explore for the rules of ECA, the set of dyadic operations hidden in each of these three-variable functions (we will see that dyadic operations implicitly include the monadic functions that, duplicated, make up the set of 16 logical functions of two variables).

The objective is to evaluate whether there is any correlation between the nature of this set of hidden functions and the dynamic class of the automaton. In a further step, this idea was extended to the interactions responsible for the second generation of ECA cells. This sought to see how the computational capabilities of the automaton can change when the temporal evolution has already begun. Therefore, as we will show in the next section, we determine two logical spectra for each rule.

9.2 Logic spectra of elementary cellular automata

We define first-order spectra as consisting of sets of dyadic logic gates that arise by filtering the three-variable logic functions of the ECA, in a manner similar to how the Fredkin gate is processed. This is done by setting one of the three variables of the rules of ECA to 1 (or vector s) or 0 (or vector n). Second-order spectra are constructed by evaluating the dyadic functions that arise in the second generation of the automaton.

Let us first partition the set of 16 dyadic functions according to the laws of symmetry conjugation, reflection, and conjugation-reflection combined, similar to Wolfram's rules. The result is shown in Table 9.1, where dyadic operations are divided into seven groups, G1 to G7. This table defines the groups, and for each dyadic operation shows the associated truth table and the first-degree Kronecker polynomials, left and right, associated with each function.

Table 9.1 shows the following groups: G1 (Mon) includes monotonic result functions that are a dyadic version of the operators K and M, which is evidenced by the

Table 9.1: The 16 dyadic matrix functions represented by the corresponding right and left Kronecker polynomials, grouped into the 7 classes that arise by applying the operations of symmetry conjugation, reflection, and conjugation-reflection combined, similar to Wolfram's rules.

Groups	Kronecker polynomials	Name	Truth tables			
			1 1	**1 0**	**0 1**	**0 0**
G1	[KK) (KK]	Mon	1	1	1	1
	[MM) (MM]		0	0	0	0
G2	[I I) (KM]	Ide	1	1	0	0
	[KM) (I I]		1	0	1	0
G3	[NN) (MK]	Neg	0	0	1	1
	[MK) (NN]		0	1	0	1
G4	[KI) (KI]	D	1	1	1	0
	[IM) (IM]	C	1	0	0	0
G5	[KN) (IK]	L	1	0	1	1
	[IK) (KN]		1	1	0	1
	[MI) (NM]		0	1	0	0
	[NM) (MI]		0	0	1	0
G6	[NK) (NK]	S	0	1	1	1
	[MN) (MN]	P	0	0	0	1
G7	[IN) (IN]	E	1	0	0	1
	[NI) (NI]	X	0	1	1	0

polynomial representation; G2 (Ide) has two elements and each of them defines an identity function since one of the variables is fixed, a fact that is also reflected in the Kronecker polynomials; G3 (Neg) is a negation also in two versions and with an operation analogous to Ide; G4 to G7 represents basic dyadic operators. As we can see, the set of 16 functions is reduced to a set of 7 symmetry groups.

To interpret the spectra that we will obtain, which will consist of sets of dyadic logical functions, it is interesting to try to measure the computational capacity of these functions. We are going to represent them symbolically by matrix operators $Op(u \otimes v)$. One option is to try to evaluate their responses to variations in their inputs. This can give an idea of the ability to transfer configurations during the temporal evolution of an automaton. Going to the extreme, it is clear that the functions of the G1 group block the transmission of any input configuration. Let us assess the overall transmission capacity of a dyadic function using the following expression Val that measures the value of the global sensitivity to changes (Mizraji 2004):

$$\text{Val}[Op(u,v)] = \frac{\partial Op(u \otimes s)}{\partial u} + \frac{\partial Op(u \otimes n)}{\partial u} + \frac{\partial Op(s \otimes v)}{\partial v} + \frac{\partial Op(n \otimes v)}{\partial v} . \quad (9.3)$$

Representing the dyadic function by its two polynomials

$$Op = U \otimes s^T + U' \otimes n^T = s^T \otimes V + n^T \otimes V',$$

results in

$$\text{Val}[Op(u,v)] = \frac{\partial Uu}{\partial u} + \frac{\partial U'u}{\partial u} + \frac{\partial Vv}{\partial v} + \frac{\partial V'v}{\partial v}.$$

This sum of derivatives from monadic operators will contain the core of our valuation. If we analyze implication, $L = [KN) = (IK]$ we have

$$\text{Val}(L) = n + s + s + n.$$

If we want a numerical version, then project the result on s and define $T[Op] = s^T\text{Val}[Op]$. In this case, we have $T(L) = 2$. Applying this evaluation to all dyadic functions, and referring it to the groups in Table 9.1, we find:

$$T(G1) = 0; \ T(G2 \ to \ G6) = 2; \ T(G7) = 4.$$

9.3 First-order spectra

These spectra arise from dissecting the rules of the ECA by means of the following six Boolean inputs: $(1xy)$, $(x1y)$, $(xy1)$, $(0xy)$, $(x0y)$, $(xy0)$. In general, we will express the inputs as vectors because the calculations arise easily from the matrices representing the second-degree Kronecker polynomials, as we will show later. If we apply the vector version of these inputs to Wolfram rules 202 and 172 that resulted from the selection of outputs 2 and 3 of the Fredkin gate, we get the following spectra:

Rule 202: D, $[NM]$, L, C, $[I\,I]$, $[KM)$
Rule 172: L, C, D, $[NM]$, $[KM]$, $[I\,I)$

Here appear all the dyadic functions that Fredkin's gate had latent. Negation, as well as identity, are implicit in these functions and arise when a second variable is fixed. This is confirmed by the following calculation:

$$[I\ I)(u \otimes s) = (I \otimes s^T + I \otimes n^T)(u \otimes s) = Iu$$

$$[NM)(u \otimes s) = (N \otimes s^T + M \otimes n^T)(u \otimes s) = Nu.$$

9.4 Second-order spectra

To explain what is sought in this second-order spectra, let us look at the following diagram:

$$
\begin{array}{ccccc}
x & a & b & c & y \\
 & a' & b' & c' & \\
 & & b'' & &
\end{array}
$$

Here, $(x\,a\,b\,c\,y)$ represents the cells of a region of the automaton, centered on b, in time $T = 0$. Let us call the rule of that automaton f. Then for $T = 1$ the state b' it is given by $b' = f(a, b, c)$. For the second generation, in $T = 2$, the state b'' is given by $b'' = f(a', b', c')$. If we now develop this equation, it turns out that

$$
b'' = f(f(x, a, b), f(a, b, c), f(b, c, y)). \tag{9.4}
$$

If we fix the central variables, for example, $abc = 101$, we get a function of x and y, $g_{101}(x, y)$. This is a dyadic function that expresses how the state of the central cell is influenced by states x, y when the central cells are fixed at 101. If we assume that dyadic functions reflect computational capabilities, we will evaluate how those computational capacities are altered in $T = 2$. The second-order spectrum is the set of functions

$$
b'' = g_{abc}(x, y) \tag{9.5}
$$

where each function is evaluated for the eight possible values of the triplet (abc). Note that these functions $g_{abc}(x, y)$ assess the functional engagement between these distant variables and can be a good measure of whether the automaton's law has the ability to transmit or block complex configurations. We will now look at explicit formulas for calculating both types of spectrum (Mizraji 2006).

9.5 Equations for the calculation of spectra

9.5.1 First-order spectra

First-order spectra are obtained from the right and middle Kronecker polynomials of matrix A (that represents the automaton's law in matrix format). Thus, the first equation that we show below arises from developing the expression $A(s \otimes u \otimes v) = A(s \otimes I \otimes I)(1 \otimes u \otimes v) = A(s \otimes I \otimes I)(u \otimes v)$. Note that, according to the fourth property of the Kronecker product shown in Chapter 2, s is a Qx1 column vector multiplying 1, which can be considered a 1x1 matrix. In this way, representing A using the associated second-degree Kronecker polynomial, $A = (W_1 W_2 W_3 W_4)$, the terms with n^T at the left disappear and only the two monadic matrices W_1 and W_2 remain (see the details of these polynomials in Chapter 4). The argument is similar for the input $(0xy)$. For the other cases, we use the right polynomial $A = [U_1U_2U_3U_4)$. The resulting equations are:

Input $(1xy)$: $Z = [W_1W_2)$;
Input $(x1y)$: $Z = [U_1U_2)$;
Input $(xy1)$: $Z = [U_1U_3)$;
Input $(0xy)$: $Z = [W_3W_4)$;
Input $(x0y)$: $Z = [U_3U_4)$;
Input $(xy0)$: $Z = [U_2U_4)$.

9.5.2 Second-order spectra

To explicitly determine the equations that allow the calculation of second-order spectra, we use the structure of the matrix A that encodes each Wolfram rule, now considering only the right and left polynomials:

$$A = [U_1U_2U_3U_4) = (V_1V_2V_3V_4].$$

Let us assume that the central triad is $abc = 100$. So, for the first generation $(a'b'c')$ the following equations result:

$$a' = A(u \otimes s \otimes n) = U_2u,$$

$$b' = A(s \otimes n \otimes n) = U_4s,$$

$$c' = A(n \otimes n \otimes v) = V_4v.$$

Here, the parameters a,b,c are the Q-dimensional vectors that make up the central triad and u,v are the variables we focus on. The second generation of b is given by

$$b'' = A(U_2u \otimes U_4s \otimes V_4v) = B(u \otimes v).$$

Note that the three variables that are to be processed by A can be expressed as

$$(U_2u \otimes U_4 s \otimes V_4v) = (U_2 \otimes U_4 \otimes V_4)(I \otimes s \otimes I)(u \otimes 1 \otimes v) =$$

$$(U_2 \otimes U_4s \otimes V_4)(u \otimes v).$$

Consequently, we have

$$[A(U_2 \otimes U_4s \otimes V_4)](u \otimes v) = B(u \otimes v).$$

In this way, the dyadic matrix B corresponding to this central triple is

$$B = [A(U_2 \otimes U_4s \otimes V_4)].$$

From the detailed analysis of these expressions for the different triples, the following equations emerge:

Case 1: (*abc*) = (111).
If $U_1 s = s$, then

$$B = U_1 U_1 \otimes s^T V_1 + U_2 U_1 \otimes n^T V_1.$$

If $U_1 s = n$, then

$$B = U_3 U_1 \otimes s^T V_1 + U_4 U_1 \otimes n^T V_1.$$

Case 2: (*abc*) = (110).
If $U_2 s = s$, then

$$B = U_1 U_1 \otimes s^T V_2 + U_2 U_1 \otimes n^T V_2.$$

If $U_2 s = n$, then

$$B = U_3 U_1 \otimes s^T V_2 + U_4 U_1 \otimes n^T V_2.$$

Case 3: (*abc*) = (101).
Yes $U_3 s = s$, then

$$B = U_1 U_2 \otimes s^T V_3 + U_2 U_2 \otimes n^T V_3.$$

If $U_3 s = n$, then

$$B = U_3 U_2 \otimes s^T V_3 + U_4 U_2 \otimes n^T V_3.$$

Case 4: (*abc*) = (100).
If $U_4 s = s$, then

$$B = U_1 U_2 \otimes s^T V_4 + U_2 U_2 \otimes n^T V_4.$$

If $U_4 s = n$, then

$$B = U_3 U_2 \otimes s^T V_4 + U_4 U_2 \otimes n^T V_4.$$

Case 5: (*abc*) = (011).
If $U_1 n = s$, then

$$B = U_1 U_3 \otimes s^T V_1 + U_2 U_3 \otimes n^T V_1.$$

If $U_1 n = n$, then

$$B = U_3 U_3 \otimes s^T V_1 + U_4 U_3 \otimes n^T V_1.$$

Case 6: $(abc) = (010)$.
If $U_2 n = s$, then

$$B = U_1 U_3 \otimes s^T V_2 + U_2 U_3 \otimes n^T V_2.$$

If $U_2 n = n$, then

$$B = U_3 U_3 \otimes s^T V_2 + U_4 U_3 \otimes n^T V_2.$$

Case 7: $(abc) = (001)$.
If $U_3 n = n$, then

$$B = U_1 U_4 \otimes s^T V_3 + U_2 U_4 \otimes n^T V_3.$$

If $U_3 n = s$, then

$$B = U_3 U_4 \otimes s^T V_3 + U_4 U_4 \otimes n^T V_3.$$

Case 8: $(abc) = (000)$.
If $U_4 n = s$, then

$$B = U_1 U_4 \otimes s^T V_4 + U_2 U_4 \otimes n^T V_4.$$

If $U_4 n = n$, then

$$B = U_3 U_4 \otimes s^T V_4 + U_4 U_4 \otimes n^T V_4.$$

Applying these equations, let us see what the first- and second-order spectra are for three Wolfram rules with very complex dynamics. We show first their three second-degree Kronecker polynomials and then the associated spectra.

Rule 30: $A(30) = [NNNI) = (MNKI) = (MNKI]$
Spectrum 1: $[NN), X, [NN), X, P, D$
Spectrum 2: $[NN), X, [I\,I), [I\,I), [I\,I), [I\,I), [NN), X$

Rule 45: $A(45) = [NNIN) = (NMIK) = (MIKN]$
Spectrum 1: $[NN), E, X, [NN), [NM), [KN)$
Spectrum 2: $[I\,I), X, [I\,I), E, [NN), [NN), [NN), [I\,I)$

Rule 110: $A(110) = [NKKM) = (NIKI) = (NIKI]$
Spectrum 1: $S, [KM), S, [KM), X, D$
Spectrum 2: $[MK), [IK), [MM), [KM), [KM), [KM), [KK), [KM)$

Since it is not easy to reach conclusions by looking at the structure of the spectra, we can adopt a notation that refers to the seven classes in Table 9.1 and that contributes to obtaining some general notions. In this alternative representation of the spectra, the distribution of dyadic functions within each class (G1 to G7) will be described for each spectrum. For this representation, we adopt the following notation:

$$S1(R) = a_1 \quad a_2 \quad a_3 \quad a_4 \quad a_5 \quad a_6 \quad a_7 \tag{9.6}$$

$$S2(R) = b_1 \quad b_2 \quad b_3 \quad b_4 \quad b_5 \quad b_6 \quad b_7 \tag{9.7}$$

$S1(R)$ represents the spectrum of order 1 of the rule R and each a_i indicates how many times class i is represented in the spectrum. For $S2(R)$ and b_i we use the same criteria described for first-order spectra. The spectra for all 88 Wolfram rule classes have been published (Mizraji 2004, 2006). Here we will only show representative cases of the four dynamic classes of Wolfram.

Class 1. Simple ECA
$S1(8) = 3001200; S2(8) = 8000000$
$S1(136) = 2202000; S2(136) = 6200000$

Class 2. Periodic ECA
$S1(23) = 0000060; S2(23) = 6002000$
$S1(58) = 0111210; S2(58) = 3211100$

Class 3. Chaotic ECA
$S1(30) = 0021012; S2(30) = 0420002$
$S1(45) = 0020202; S2(45) = 0330002$

Class 4. ECA with confined complexity
$S1(110) = 0201021; S2(110) = 2320100$

Let us now summarize the conclusions of the study of the total spectra (Mizraji 2006). The main conclusions that we will mention allude to trends, but there are no necessary or sufficient conditions to foresee a dynamic:

1. The propensity to generate complex dynamics is usually associated with the presence of the G6 or G7 functions (and eventually the G2–G3 pair) in the S1 spectrum and the persistence of G7 or the G2–G3 pair in the S2 spectrum.
2. The influence of G7, G6, and the G2–G3 pair on both spectra depends on the amount of G1. A high amount of G1 in the spectra indicates a tendency to block transmission along T of complex configurations.
3. Rule 30 is a case in which there is an indicator of a trend towards complexity signed by G7 in S1 together with G7 and the pair G2–G3 in S2, and the absence of G1 in both spectra. The situation is similar in rule 45, which is also chaotic.
4. Rule 110 does not show G1 and does have G6 and G7 in S1, but in S2 there are two representatives of G1 and also the pair G2–G3.

Let us point out that G2–G3 together represent a coupling of the dyadic versions of identity and negation, which in their monadic version define the G7 group.

9.6 Hidden potentialities

The emergence of complex configurations or events in the natural world, including human cognition, its social organization, its technology and its science, has in many cases been difficult to explain from classical conceptions of science. This led, as is usual in areas where knowledge is incomplete, to the creation of metaphysical positions such as extreme reductionism or holism, or metaphysical emergentism, and at the basis of these conceptions, there is always the conversion of a transitory ignorance into an immovable philosophical position. Thus, an aphorism such as "the whole is more than the sum of the parts" went from being a reasonable statement to imposing itself as a postulate of impossibility: "the properties of water can never be understood from the isolated properties of oxygen and the hydrogens that compose it." We can counter, perhaps also dogmatizing, that a statement of this kind is asymptotically false. And the asymptote will be reached when our science understands what is in the properties of those atoms that predispose them to combine into molecules with the potentialities of forming a liquid like water.

At the beginning of his book *Life, A user's Manual*, Georges Perec, referring to the pieces of a puzzle, challenges us with the following comment: "[. . .] the parts do not determine the pattern, but the pattern determines the parts: knowledge of the pattern and of its laws, of the set and its structure, could not possibly be derived from discrete knowledge of the elements that compose it." This phrase seems like a holistic manifesto but has, as was Perec's habit, a profound subtlety. Thinking about the parts and the whole, or about water and its atoms, let us establish a small axiom and see its consequences: "If Y is built from X, then we call X an Atom." So we ask ourselves, what is the "atom" of the puzzle, the small piece or the whole of the drawing? The answer promoted by the axiom is clear: here the atomic unit is the whole image, and the emergent produced by the initial large image and the intelligence of the craftsman is the small sculpted piece. Continuing with the analogies suggested by Perec's text, let us think that the "atom" of a story created by authors who carefully organized their stories is not in the words or the episodes, but is the project that guided the creation of the story. Human beings are basically designers, and the structure of a motor cannot be induced from the structure of a screw; the motor design is the atom X from which the screw comes out as the piece of the puzzle.

But liquid water is not a project that precedes hydrogen and oxygen atoms. A squid's eye is not a project that precedes the primordial cells that after millions of years of evolution led to the squid and its eye. Our need to create concepts is not prior to the creation of the networks of nerve cells that make up our brains. In his great film "Alphaville" in 1965, director Jean-Luc Godard created the Alpha 60 computer, which mercilessly ruled that mysterious city. While detective Lemmy Caution advances in the investigation for which he was hired, Alpha 60 emits as a sound background to his iniquities, phrases and thoughts stolen by her from multiple famous authors. Experts have entertained themselves by looking for these authors; in this way they detected phrases by Pascal, Nietzsche, Baudelaire, and, abundantly, Borges. I now transcribe an interesting phrase from those stolen by Alpha 60: "Once we know the number one, we believe that we know the number two, because one plus one equals two. We forget that first we must know the meaning of plus." This phrase from Alphaville puts the scientific problem of the emergency in fair terms. It does not assume that it is impossible to understand, but it points out what is missing to understand. Mathematics long ago clarified the meaning of "plus" and for standard arithmetic, $1+1=2$ He does not seem to keep secrets. But taken to other scales, this metaphor is still valid.

All this leads us to ask whether the computational irreducibility assigned to the dynamics of some automata is a final, demonstrable property, or whether it is a transient state of our knowledge.

In ECA universe, the emergence of complexity during the temporal evolution, or the impossibility of sustaining complexity, depends on latent properties in the law of automata. These elementary models have the power to provide us with novel dynamic situations that lead to reflection regarding their possible extension to other situations. Here we will show two cases that exhibit latent potentialities: Rule 120 and Rule 164 (Arruti and Mizraji 2006). We will first show the experimental results that arise from computing these automata. Then we will see the structure of the logical law that governs each case and the associated spectra.

9.6.1 Rule 120

When exploring this rule for random initial conditions, but with a low probability of living cells, a very interesting phenomenon appears. For these computations, we use Simone Maggi's WAUTOM program, where the live cells are white and the empty cells are black. In this automaton, one-dimensional space is annular, so configurations that disappear on one flank in T appear on the other $T+1$. Figure 9.1 shows three evolutions of automaton 120. The images were chosen from many experiments and deliberately arranged. This is an automaton belonging to Wolfram's class 4. Here, we use $p = 0.05$ (which is the average with which each living cell is drawn in $T=0$; therefore, the effective number varies in each draw). The interesting phenomenon we mentioned is the following. If the living cells are highly isolated as in panel A of the figure, these living cells remain isolated and show no signs of complexity. In panel B, nearby living cells appear in the upper right corner of the diagram, and we can already see how the neighborhood generates an interaction that begins to produce complex configurations. And in panel C, the randomness of the initial draw produced more living cells and more adjacencies, showing the emergence of an exuberant complexity that is exacerbated by the creation of new close neighbors.

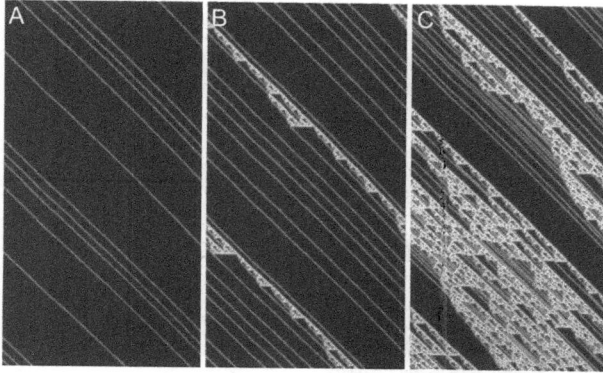

Figure 9.1: Rule 120 automaton evolution for random initial conditions with a probability of living cells $p = 0.05$. This automaton displays drastically different behavior depending on the adjacency of living cells in the initial condition, where its behavior shifts from extreme simplicity to extreme complexity as the number of living cells increases. Years after the publication analyzed here, this rule was reclassified and incorporated into category 4, the most computationally rich rule.

Rule 120 corresponds to a Kronecker polynomial, which in turn corresponds to two logical operations of simple matrix expression:

$$W(120) = [NIII) = E(I \otimes S) = X(I \otimes C).$$

First-order spectrum:

$$X, \ [I\,I), \ X, \ [I\,I), \ P, \ C.$$

Second-order spectrum:

$$[NN), \ [I\,I), \ X, \ [I\,I), \ E, \ [I\,I), \ [I\,I), \ [I\,I).$$

Encoding these spectra using expressions (9.6) and (9.7) results in:

$$S1(120) = 0 \ \ 2\,0 \ \ 1 \ \ 0 \ \ 1 \ \ 2$$

$$S2(120) = 0 \ \ 5 \ \ 1 \ \ 0 \ \ 0 \ \ 0 \ \ 2$$

9.6.2 Rule 164

In the experiment with which we illustrate this rule, we start from a situation opposite to the previous one. Here we establish the draw with a very high probability (0.95) that in $T = 0$ a cell is alive. After this condition is applied repeatedly, we select, as shown in Figure 9.2, two situations with different end states.

Its polynomial representation is $W(164) = [I\,N\,I\,M)$. Let us look at the spectra corresponding to this rule.

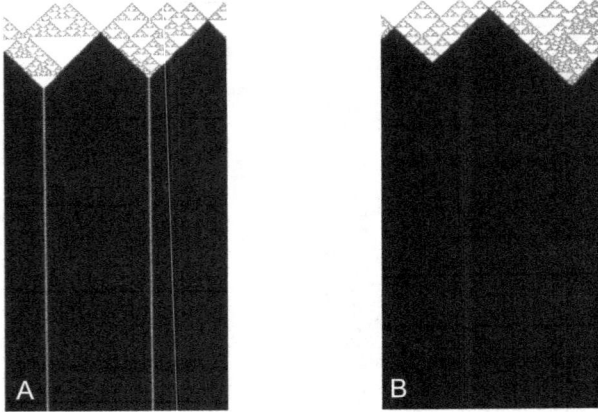

Figure 9.2: Evolution of automaton 164. The initial conditions are drawn with a $p = 0.95$ of active cells in the initial state. This rule has a leafy beginning, suggesting a potentially complex evolution, but it is frustrated and ends with a complete collapse of that potential. As analyzed in the text, this automaton adds blocking signatures to the complexity signatures.

First-order spectrum:

$$E, \ C, \ [I\,I), \ [M\,I), \ [KM), \ [NM).$$

Second-order spectrum:

$$E, \ C, \ E, \ [MM), \ C, \ E, \ [MM), \ [MM).$$

The coded version of these spectra is:

$$S1(164) = 0 \ \ 2\,0 \ \ 1 \ \ 2 \ \ 0 \ \ 1.$$

$$S2(164) = 3 \ \ 0 \ \ 0 \ \ 2 \ \ 0 \ \ 0 \ \ 3.$$

The automata $W(120)$ and $W(164)$ show in their dynamic behavior, two kinds of hidden potentialities. $W(120)$ shows the potential to possess complexity only if associated active cells appear. On the contrary, $W(164)$ shows that a strong adjacency of active cells initially begins to trace complex configurations, but then there is a total disappearance of complex structures and even of active cells. Let us see what the spectra say in each case:

1) In the case of rule 120, we see in $S1(120)$ complexity indicators such as the presence of logical functions of groups G6 and G7 and the absence of blocking functions G1. On the contrary, the second-order spectrum $S2(120)$ once again shows functions of the G7 group and the absence of G1 blockers. Consequently, both spectra have signs of prospective complexity. But the relevant fact of this rule 120 (which, as we said, belongs to Wolfram's class 4) is that latent complexity does not arise in isolated living cells, as Figure 9.1 A shows.

2)   In the case of rule 164 (Wolfram's class 2), we also see interesting behaviors and data: The leafy start when the active cells are very close, and then a sudden and massive collapse. If we analyze the structure of the spectra, we find suggestive explanations for the initial leafiness and for the subsequent collapse. The $S1(164)$ spectrum shows complexity indicators such as the presence of the G7 group together with the absence of elements of the G1 group. But the interaction spectrum $S2(164)$ shows a totally different situation, where the G7 class remains increased but with a strong presence (three in eight) of elements of the G1 block group.

These small three-variable automata suggest explanatory possibilities for various biological and social events. In the case of rule 120, we see that if an isolated or very thin set of individuals exists at the initial states, we would not be able to detect the hidden complexity that the automaton possesses, since this only becomes evident during the evolution of "socialized" cellular individuals. Let us now move into the world of conjecture. There are many situations suggested by this dynamic, where complexity emerges only under certain conditions. During human evolution, socialization awakened innumerable hidden potentialities that the isolated individual was not able to exhibit previously. Some time ago, anthropologists located the mysterious appearance of human language to only about 40,000 years ago, when our species had already been on the Earth for about 200,000 years and probably possessed, from the beginning, the larynx, the tongue, and the brain necessary to produce articulate language. It was argued that about 40,000 years ago, the climate imposed socialization, which also involved famous creations in cave art. Today, this "social" hypothesis is questioned although not refuted, and it is a question of seeking a genetic and molecular basis for the linguistic innovation that *Homo sapiens* exhibits compared to other primates. In any case, let us evoke the many technological and scientific developments that humans achieved after their socialization, such as the creation of agriculture, the invention of writing, the innumerable technological inventions, the progress of medicine, the rapid development of aviation, or the evolution of computer languages, among many others. In the example of rule 164, we see that a leafy socialization can begin by generating a great wealth of configurations in the initial times, and then suddenly a few cells survive or become completely extinct. This is also suggestive of biological, social or economic events, where after a splendid beginning, systems suddenly collapse. The hidden causes of the spectacular initial deployment of certain systems (sometimes brilliant people of promising destiny, or entrepreneurial initiatives, or scientific theories), which seem destined to be overwhelmingly successful but suffer a sudden collapse, are generally enigmatic, but the dynamic behavior of a miniature like the automaton 164 can be a source of ideas.

Obviously, none of the complex situations emerging in the real world can be explained by ECA. All the explanations that these automata suggest will have to be framed and investigated within the logic and procedures that operate in the field of investigation of the situation that is sought to be resolved. But as research instruments, these cellular automata have an extremely powerful virtue: it is their ability to introduce subtle concepts and innovative ideas into the researcher's mind, concepts and ideas that could be inaccessible to the researcher immersed in the extensive and tangled amount of facts and links with which reality surrounds them.

References

Arruti, C., Mizraji, E. (2006) Hidden potentialities, International Journal of General Systems, 35: 461–469.

Feynman, R.P. (1999) Lectures on Computation, Perseus, Cambridge.

Fredkin, E., Toffoli, T. (1982) Conservative logic, Intern, Journal of Theoretical Physics, 21: 219–253.

Mizraji, E. (1996) The operators of vector logic, Mathematical Logic Quarterly, 42: 27–40.

Mizraji, E. (2004) The emergence of dynamical complexity: An exploration using elementary cellular automata, Complexity, 9(6): 33–42.

Mizraji, E. (2006) The parts and the whole: Inquiring how the interaction of simple subsystems generates a complex system, International Journal of General Systems, 35: 395–415.

Index

https://doi.org/10.1515/9783112230053-011